The Climate Change Illusion

And the Real Causes of Global Warming

by Edward Rouse Pryor

Published by

HHH

Hanway Hopcott House

P.O.F.B.

The Climate Change Illusion

For information about this title or to order other books and/or electronic media, contact the publisher:

Hanway Hopcott House
7740 Indian Oaks Drive
Department G-301
Vero Beach, FL 32966
hanwayhopcotthouse@gmail.com

ISBN: 978-1-7341632-0-9 (paperback)
978-1-7341632-1-6 (ebook)

Printed in the United States of America

Dedicated to:

Beverly J. Pryor

for her selfless and unwavering support,

encouragement, and enthusiasm.

Contents

The Climate Change Illusion

And the Real Causes of Global Warming

Who wants to understand the poem, must go to the land of poetry.

— Johann Wolfgang von Goethe, 1819.

Foreword

THE THESIS OF *THE Climate Change Illusion* is that – "yes, the earth is warming – no, the warming is not being caused by humans or greenhouse gases". Recently identified magnetic forces from the sun which are affecting the earth's sustained cloud cover and the amount of solar heat reflected away from the earth are the principal causes of current global warming and climate change. The newly discovered solar "dual-magnetic wave" cycle takes several hundred years to go from a minimum global temperature to a maximum. We are approaching the temperature maximum of the current cycle and within less than a decade should slowly start the downturn toward the next minimum – projected to be in another three to four hundred years.

The well-documented "Little Ice Age" in the late 1600s when the Thames River in London froze solid and "Frost Festivals" were celebrated, occurred about 370 years ago at the beginning of the current solar dual-magnetic wave cycle. The world is currently experiencing the rising temperature that results at the other end of this natural cycle.

The Climate Change Illusion, clearly and engagingly takes the reader into the scientific basis both for it not being greenhouse gases that are causing our warming, and also for solar magnetic forces that influence earth's varying cloud cover being the controlling factors. The book makes a compelling case while being amazingly readable.

Finally, the scientific community is beginning to understand the real causes of global warming and climate change.

— Robert J. Cotter, (PhD in Chemistry - M. I. T.)

Preface

THIS FASCINATING BOOK COMPLETELY turns the understanding of the world of climate change upside down.

It explains why changing atmospheric greenhouse gas content of the earth's atmosphere does not have enough forcing power to cause the global surface temperature changes of consequence we have seen and are seeing.

It then clearly explains how the sun does indeed have the forcing power to cause the changing global surface temperatures we have witnessed and are witnessing for short-term, mid-term, and long-term temperature changes and why each has a different cause.

This is an enlightening and compelling book with a message of astonishing magnitude.

Introduction

THE EARTH'S SURFACE IS warming – there's no doubt about it. But what's causing that warming? Is it the human-caused emission of greenhouse gases into the atmosphere – a potential cause that is championed by atmospheric-oriented climate scientists – or is it natural causes that are less publicly well known (but are becoming understood and characterized by solar oriented climate scientists)? If the cause is humankind's emission of greenhouse gases into the atmosphere, the prognosticated warming could go on until the earth's fossil fuels are depleted – and if some predictions turn out to be true, the world could get a lot hotter than now. This last possibility has resulted in ominous federal reports[1] and a legion of peripheral "scientific apparitionists" to project a world of inundating seas, failed crops, economic collapse, an atmosphere too hot for humans to breathe, intolerable floods and wildfires, and long-dormant plagues becoming freshly revived[2]. Threaded through these disaster stories is the promise that if we just stop using those fossil fuels and stop emitting greenhouse gases like carbon dioxide into the atmosphere – and substitute wind generators and solar panels, humankind will be all set for a rosy future.

All the above cataclysmic scenarios are based on a scientifically unconfirmed premise or unestablished supposition: changes in earth's atmospheric carbon dioxide (CO_2) content cause consequential changes in global surface temperature. Could this widely accepted, but actually speculative supposition be incorrect? If it is invalid, then the whole projected series of devastating scenarios is nothing more than imaginative conjecture – a series of pipe dreams. There are recent accumulating indications that as the anthropogenic warming "movement" was building and as these imagined scenarios of climate disaster were unfolding, the evidence that increasing atmospheric greenhouse gases was causing consequential global surface temperature increases was beginning to fall apart.

So let's dig a little deeper into this premise that humankind's emission of greenhouse gases into the atmosphere is the principal cause of global warming and climate change. Perhaps we'll find it is not as straightforward as commonly thought.

Part I
The Rise and Fall of the Anthropogenic Premise

For a scientist, skepticism is the highest of duties', blind faith the one unpardonable sin…

—Thomas H. Huxley, 1866

Chapter 1

Confusion

Anyone who isn't confused doesn't really understand the situation.

— Edward R. Murrow c 1964

The earth is currently getting warmer – and it has been doing so in bits and spurts for the last 370 (or so) years. We know this because, aside from other proxy indicators (like the thickness of tree rings and the trace contents of air bubbles in ice cores), we can measure recent temperatures with our relatively modern instrument – the thermometer (see graphic 1.1).

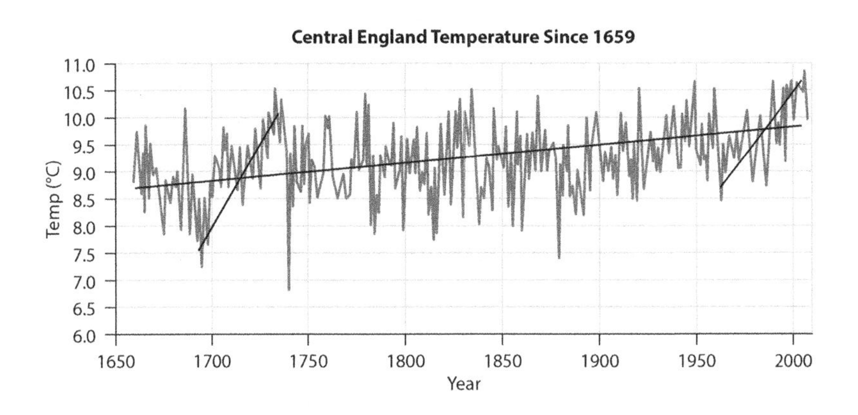

Graphic 1.1. **The Midlands of England** (NCAR/after G. Manley)[3]. This three and a half century chronicle is the longest reliable instrument (thermometer) recording of temperature in the world. [Time →].

The remarkable temperature history shown in graphic 1.1, tells us there has been a modest continuous temperature increase over the past three and a half centuries while there have also been shorter-term (half century) bursts of temperature increase (as well as short-term periods of declining temperature). This has been the case ever since the Little Ice Age in the late 1600s. An inner core of climate scientists at around the turn of the twenty-first century worried about the half-century short-term temperature rise from about the 1960s to the early 2000s, shown by the right-side short-uptick temperature trend line of graphic 1.1. They said they only could explain this temperature rise by including man's use of carbon fuels in their computer climate models (which only went back as far as the late 1800's). But we must also consider the similar short-uptick rise from the 1680's to the 1730's – before the computer model data began and before the Industrial Revolution began in the late 1700s with its ever-increasing emissions of anthropogenic (human-caused) carbon dioxide into the atmosphere. The short-term trend during this period is more extensive and increases faster than the right-side short-term rise. If the right-side (recent) uptick only can be explained by human-caused increasing atmospheric carbon dioxide, what explains the uptick of the 1600s? Did the inner-core of climate scientists not look back as far as they should have when they formulated their models and drew their conclusion? Have they become too worried by a recent short-term temperature increase that is not at all unusual?

By way of comparison let's look at the earth's temperature over the last 420,000 years (graphic 1.2).

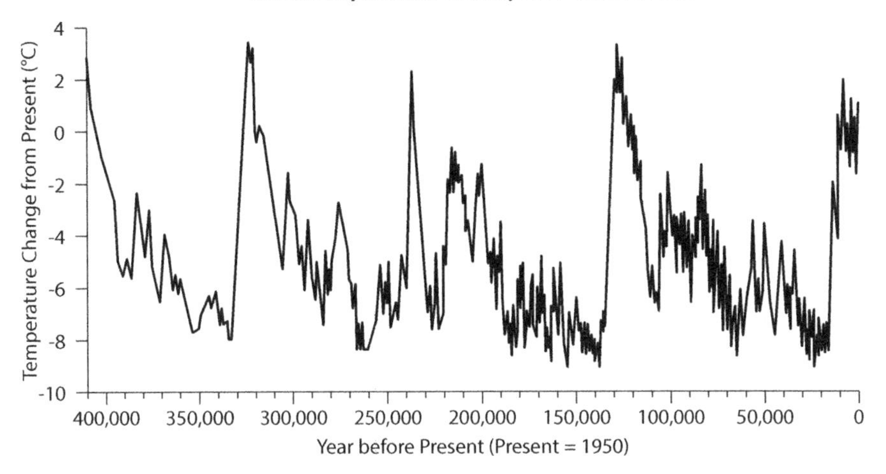

Graphic 1.2. Global Temperature Change Over 420,000 years (NOAA Paleoclimate/after Petit et al)[4]. [Time →].

Graphic 1.2 is a fairly reliable measure based on well-dated climate proxies from the Antarctic Vostok deep ice-core. You will notice about a 12°C excursion from the peak of the various genial (or elevated-temperature, inter-glacial) stages to the hostile valley of the lower-temperature, glacial stages[5]. What causes these temperature excursions? Note that in the current genial stage (where we are now – to the right) the peak is not as high as the previous cycle peak some 100,000 or so years ago, but we're almost as high as the peak before that. So, are we naturally going to go higher on this cycle or have we already peaked? Based on the overall height of our current excursion on graphic 1.2 and on the rhythm of the peaks we would have to conclude we must be past the main-peak and somewhere near the next and perhaps final sub-peak. But is that five, or even more years away? The scales of these charts are so compressed it's hard to tell if we have already peaked or not. Thus, the temperature may continue to get a little warmer, but not a lot warmer – for even if we rise to the height of the last main-peak it would be little more than one degree Celsius higher until we go over the top and begin the normal slide down toward the next valley. As you will see later on, some European Center for Nuclear Research (CERN) scientists suggest that the peak we still seem to be ascending toward might come within less than a decade – by the mid-2020s. So perhaps the temperature rise we are seeing now is part of a perfectly normal cycle of varying global temperature based on other "natural" factors – possibly influenced by the sun. Two of the questions for us now are:

- *What are the basic causes of the earth's temperature variations?*
- *Should we be concerned over the more rapid rise in temperature over the last 60 to 70 years shown as the right-side uptick in graphic 1.1, or is that simply a normal, natural variation?*

While the central question is:

- *Is the human use of fossil fuels causing the earth to warm?*

The Causes of Global Warming

Until very recently, nearly all scientists agreed there were four leading possible causes of global warming over the last several hundred years:

- Increases in greenhouse gas content of the atmosphere (such as carbon dioxide, primarily from human use of fossil fuels).
- Changing solar heat intensity.
- Changes in earth's albedo or reflective power – primarily due to changes in cloudiness that changes the amount of solar heat reflected away from the earth.

- That great climate flywheel, the ocean, with its many temperature layers and complex colliding thermohaline major currents that bring layers of different temperature water to the surface. This redistributes heat between the ocean and the atmosphere making the apparent temperature of the earth's surface change without there being any externally influenced change in the overall heat content of the earth and its oceans and atmosphere.

(A fifth cause, the sustained orbital change in distance between earth and sun is not discussed here because its effect over several hundred years is so small as to be undetectable.)

There are other causes but they appear to be minor. For example, the amount of heat coming from the interior of the earth to the surface is estimated to be about 0.03 percent of the amount coming from the sun and thus is considered inconsequential.

All these causes have contributed to our confusion over just what is happening to the earth's surface temperature and what is causing it to happen. But one thing seems apparent – the earth's surface is warming – at least right now, and that is alarming to many.

The first of the four prime possible causes for that warming – the premise that human-caused greenhouse gas emissions (mainly carbon dioxide) are the major cause of current climate warming – is the only one of these possibilities with its own advocacy organization, which is complete with a very powerful "public awareness" (public opinion guidance) apparatus, and thus it has been given widespread attention. In fact, human emission of greenhouse gases into the atmosphere has completely dominated the conversation about global warming over the past few decades[6].

In this book, we readily acknowledge that the earth's surface temperature is indeed currently warming and also that atmospheric CO_2 content is rising. So it might seem to some people that one is causing the other. But two things happening at the same time does not necessarily mean that one is causing the other. The correlation could be entirely coincidental and each of these factors could be driven by completely independent causes. We will dig deeper into this matter. And we will evaluate and discuss the relative importance of all potential causes of the current short-term, as well as mid-term, and long-term global temperature variation. Then we will attempt to determine just how much each suspected cause actually contributes to the warming and cooling we've witnessed over various time frames. And we will attempt to put the recent (over the last five or six decades of the twentieth century) global temperature rise which seemed to influence the anthropogenic scientific community so much, into proper perspective.

The "Alert" and the "Pronouncement"

In June of 1988, Dr. James Hansen, a respected NASA scientist and head of the Goddard Institute for Space Studies, alerted the American people to the prospect of anthropogenic global warming by testifying before congress that the increasing human-caused atmospheric carbon dioxide (CO_2) emitted by the use of fossil fuels was already causing the earth to warm[7]. This "front-page" news began to get the attention of people around the world and caused increasing concern in the environmental community. It heightened the interest in and scientific study of the prospect of anthropogenic global warming and attendant climate change. The Intergovernmental Panel on Climate Change (IPCC) was formed later that same year (1988).

In early 2001, more than a decade later, it was reported that another one of a small group of elite atmospheric climate scientists publicly announced that they were sure "beyond doubt" that humankind and its use of fossil fuels was the dominant cause of the global warming that we had witnessed over the previous half-century or so[8]. This pronouncement was made in response to a commitment the IPCC had made a decade earlier in their 1990 First Assessment Report. They had stated it would take about a decade more to conclusively answer the question that President Jimmy Carter had posed to the National Science Foundation in 1979: *"Is the human use of fossil fuels causing the earth to warm?"*

The United States Congress had agreed it was an important question to scientifically answer since speculation was rife at the time that humankind was responsible for global warming, but there was no credible scientific explanation or validation of that notion. Furthermore, Congress had opened the financial resources of the federal government for research to scientifically answer the question and the scientific community had enthusiastically availed themselves of this open checkbook[9]. Within only a few years United States government funded scientific research on the topic of global warming or climate change reached some $2 billion per year.

As the end of the twentieth century approached, people both within government and outside of it were expecting a scientifically validated answer to the question of what the human role was in global warming. The IPCC had committed itself to provide an answer by then and the total of these research expenditures was approaching $50 billion, so there was a strong incentive for the inner core of climate scientists to announce an answer. And they did have a judgmental case that suggested that humankind and its addition of greenhouse gases to the atmosphere might be responsible, although it was far from being scientifically certain.

After conferring together, the inner core of climate scientists gave the world their answer on January 21, 2001. They knew they were the acknowledged climate experts and that other scientists would probably accept their conclusion based on the evidence they had assembled if it was properly presented. They had a seemingly credible theoretical (physics) calculation showing that a doubling of atmospheric carbon dioxide would result in a 1°C rise in global surface temperature. This would result in a water vapor feedback amplification of another 2°C, for a net increase of 3°C. In addition, they had several pieces of apparently valid observational evidence: The so-called "hockey stick" graph (graphic 3.1) that showed a correlation between recently rising atmospheric carbon dioxide and rising global temperature. They also had curves from the Vostok ice-core (graphic 1.2) data that showed four cycles of matching global temperature and atmospheric CO_2 versus time (which we will discuss later on). In addition, they had convincing computer climate model scenarios that seemed to show that humankind and our use of fossil fuels was the cause of global warming.

Perhaps most important to many in the scientific community, was the firmly held belief that changes in solar intensity were not powerful enough to be responsible (see graphic 1.3[10]).

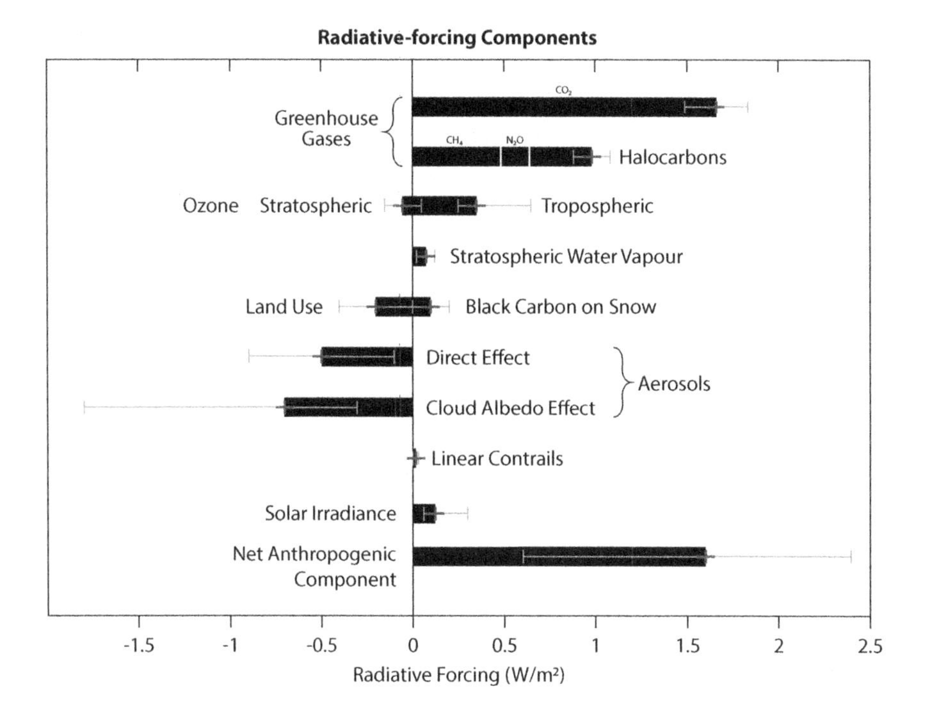

Graphic 1.3, **Radiative Forcing** (adapted from Leland McInnes, IPCC AR4)[11]. This chart is typical of the many produced over the years by the "inner core" of climate scientists. Note the tiny influence on global temperature by the sun compared to the consequential influence by greenhouse gases.

The next possible cause of the global warming we were seeing – changes in earth's albedo (or reflectivity of solar heat), that were caused by systematic sustained changes in average cloud cover – was not well enough conceptualized to be scientifically evaluated or even considered at that time. In addition, the oceanic/atmospheric "fly-wheel" influence on sustained global surface temperature was only vaguely envisioned and poorly characterized.

Thus, with all the other potential causes ruled out by lack of scientific understanding, most scientists – both inner core climate experts and those on the periphery – thought there was simply no other possible basis for the overall global warming we were witnessing than increases in atmospheric greenhouse gases (primarily carbon dioxide) that were caused by the human use of fossil fuels.

Perhaps not fully appreciating the uncertainty, and definitely not emphasizing it, the inner core of climate experts also believed that even if they were wrong in their assessment of human culpability, it was for the "common good" if the general public thought that human use of fossil fuels was responsible for global warming simply because that might be the case – and if it later turned out that was the case, and no earlier corrective action had been taken – it would be too late to do anything about it.

The time had come for the inner core of experts to announce that: "Yes, humans, with their use of fossil fuels that emitted carbon dioxide into the atmosphere, were causing the earth to warm"[12].

Most of the peripheral scientific community generally accepted this conclusion by the experts, and many peripheral scientists began to use it as a basis for proceeding on to new phases of research: How long will it be before global temperatures rise to detrimental levels? What will be the effects on the world's environment and on the human population if we don't stop using fossil fuels? Now peripheral scientists and statisticians could quantify the secondary effects of continued use of fossil fuels for such things as sea-level rise and the associated population relocation because the answer provided by the inner core included a quantification of the temperature effect of rising levels of atmospheric carbon dioxide: *a doubling of atmospheric CO_2 would cause a 3°C rise in global surface temperature*[13].

The overall amount of climate change U.S. government funding, instead of diminishing (since the answer to the Jimmy Carter question that kicked it all off some two decades earlier had now apparently been provided), began to increase – as opportunities to "end-the-madness" of the fossil fuel era took hold. Many peripheral scientists wanted to help define and hopefully prevent a global environmental disaster caused by humankind – by depicting in detail the presumed detrimental effects of continued fossil fuel use. These depictions were accompanied by a plea to curtail the use of fossil fuels. Federal government expenditures for research and development combined with financial assistance for prototype and even production scale "renewable" energy projects (generous subsidies and tax abatements mainly for wind farms and solar projects) quickly rose from $4 billion dollars per year in 1993 to $6 billion per year in 2005 to $12 billion per year in 2014 [14] (not counting private, state and local grants or the generous federal multi-billion-dollar "economic stimulus" supplements).

Some people, particularly those from within the environmentalist community, adopted a strong ideological acceptance of the human role in causing global warming. Many of them began to take on an almost evangelical zeal to spread the word and to warn the public about the dangers of continued fossil fuel use[15]. After all, it was becoming clear that humans were causing other environmental problems, so why not this one?

But despite the apparent belief by the experts, some other people – including those from the scientific, political, and general public communities – were not responding to this new-found truth for whatever reason. Perhaps natural skepticism, perhaps entrenched interest, or perhaps just not a realization of the severe consequences, were some of the possible reasons. So it became more and more incumbent upon those who "saw the light" to convert those who had not yet accepted this alarming revelation. Contrary resistance led to greater believer determination, and the subject began to evolve from quiet intellectual discussion to a matter of fierce belief – often with attendant highly polarized viewpoints.

Many scientists who were interested in the subject but not directly involved took a more traditional slightly skeptical scientific position and began to question some of the details of how the expert inner core had arrived at their anthropogenic conclusion. They were particularly interested in the details of the physics that had been so hastily offered-up and accepted as the basic cause. Up until that point there had been a number of different starting point conceptual packages with different assumptions (suppositions and postulations) formulated by different researchers for the calculations used in the physics basis. These provided a range of quite different answers for the climate sensitivity figure (the global temperature rise caused by a doubling of atmospheric CO_2), which ranged from about 0.1°C all the way to about 8°C (or even more)[16]. These climate sensitivity figures spanned the entire range from the trivial, to the consequential, to the outrageous (unbelievable) – and suddenly the experts had homed in on the consequential figure of 3°C. There were those scientists who thought this figure needed additional confirmation rather than just blind acceptance. So later in 2001 a formal National Academy of Science report stated: "*Because there is considerable uncertainty in current understanding of how the climate system varies naturally and reacts to emissions of greenhouse gasses and aerosols, current estimates of the magnitude of future warming should be regarded as tentative and subject to future adjustments (either upward or downward)*"[17].

This meant, of course, that a downward adjustment could show that the influence of carbon dioxide on climate might be taken out of the range of being consequential. But this kind of scientific caution didn't have much effect on the mainstream anthropogenic-leaning scientific, journalistic, and environmental communities who had accepted the earlier word of the experts as a fundamental truth. Accordingly, the science of climate change now became canonical as far as the broader anthropogenic-coalition was concerned – and parts of the political community joined in and accepted it as a fundamental truth (despite the National Academy of Science's cautionary statement).

So increasing atmospheric greenhouse gases being the principal cause of global warming seemed to be the final answer, at least to the loosely-aggregated anthropogenic global warming coalition – who, by now had become highly resolute. To them, the science was settled. The debate was over. People who suggested otherwise were considered malcontents and obstructionists who must have been on the payroll of carbon fuel interests and were not to be listened to. They were "deniers" whose sole objective was to cloud the "undeniable science of anthropogenic global warming" and nothing more. And this belief has pretty much continued until today.

Most of the scientific, journalistic, and environmental communities have come around to this narrative and have come to accept as a basic truth that human beings and the carbon dioxide they release into the atmosphere is responsible for the global warming we recently have witnessed. They do not seem to understand that the conclusion of the "expert" inner core was a *judgmental assessment*, rather than derived from established scientific fact. And that conclusion was primarily based on the assumption there was no other apparent plausible cause of the warming we were seeing. Any adverse suggestion about (or questioning of) that judgmental assessment (which was rapidly becoming an ironclad conventional wisdom) was termed "casting doubt on the established science" or "fostering the illusion of uncertainty". Anyone espousing such blasphemy began to be called a "*climate denier*".

Yet bits and pieces of scientific evidence that were becoming available to the scientific community were being ignored by the controlling atmospheric-oriented inner core (who were the presumed experts), apparently because these bits and pieces did not "fit" with the expert anthropogenic-cause explanation. Data of solar irradiance versus surface temperature and atmospheric CO_2 content vs surface temperature [18] sometimes showed a close correlation between solar irradiance and earth's surface temperature while also showing a wide departure between atmospheric CO_2 and temperature. Such curious observational evidence should have sent anthropogenic global warming oriented scientists scurrying to at least check it out. Yet it was simply ignored by them as they continued to blame global warming on humankind – and as they continued to research the potential effects that this postulated anthropogenic global warming would have on the world's environment.

Apparently neither the anthropogenists nor the contrarians had a real understanding of what was causing the earth to warm, but evidence like that mentioned here made it sometimes appear that it might not be increasing atmospheric carbon dioxide – something that apparently the anthropogenists simply did not want to even consider. Looking for other potential reasons for the global warming we were witnessing did not seem to interest the now-determined inner core of climate scientists. Perhaps the most alarming aspect of such behavior by the climate experts was their unwillingness to even acknowledge the possible existence of anything that departed from their human-caused atmospheric greenhouse-gas narrative. Rather than the open scientific curiosity one would expect at this stage of the scientific investigation of the topic, this narrow view would appear to be a symptom of a close-knit cult-like enrapturement – not a good place for a group of scientists to be.

A frequently used journalistic phrase was "the established science of climate change" and sometimes, to the more dramatic environmental activists, "the undeniable truth of science". But for the more skeptical scientists, those interpretations were judgmental assessments with no legitimate direct scientific basis.

The following lingering questions remained:

- *Did the level of human-generated atmospheric greenhouse gases really have a consequential effect on global surface temperature?*
- *Could something else be the cause of the global warming we were witnessing?*

Chapter 2

Early Concern about Global Warming

Follow the Flag! `

— Motto of the Wabash Railroad.

IN THE NINETEENTH CENTURY, scientists, including: de Saussure, Fourier[19], and Tyndall[20] discovered the greenhouse effect wherein something in our atmosphere (either clouds or water vapor or some of the trace greenhouse gases such as CO_2) retained heat that would otherwise escape and thus caused the surface temperature of the earth to be about 33°C higher than it would have been without an atmosphere.

At about that same time, the Industrial Revolution really got going – and coal burning steam-powered factories were sprouting up – mainly in Britain and Europe, which were emitting tons of carbon dioxide into the atmosphere. This begged the question: *If CO_2 was a consequential component of the greenhouse effect, what would be the influence of all this additional human-caused CO_2 on the temperature of the earth?* This led scientists to try to determine the "partitioning" of the greenhouse effect among its probable causes – clouds, water vapor, carbon dioxide, and the other trace greenhouse gases such as methane, nitrous oxide, and ozone, that were in the "open sky" part of the atmosphere.

Let's postulate for a moment that clouds are responsible for 20 percent of the greenhouse effect and that 80 percent of the effect comes from greenhouse gases such as water vapor and CO_2 that are present in the "open sky". And let's assume that we can further partition the open-sky effect to, say, 60 percent water vapor, 26 percent carbon dioxide, and 14 percent other trace greenhouse gases. Thus, with relatively simple arithmetic, scientists can calculate that a doubling of CO_2 in the atmosphere from pre-industrial days will cause a 1°C rise in global surface temperature. And, using more assumptions, scientists can calculate that this 1°C rise will cause enough water vapor to evaporate from the world's oceans and "green" land areas to produce a positive feedback loop and create another 2°C rise in global surface temperature for a total rise of 3°C. This is what the inner core of climatologists had done. They had answered a long-standing scientific question using an apparently straight-forward approach. This quantified answer came just at the end of the twentieth century.[21] It was proclaimed to be true by a consensus of the expert inner core of climatologists. This proclamation was accepted by the peripheral tier of scientists because they had a preconceived notion that this was correct, could see no other possible cause, and it was apparently the confirming conclusion of the "experts". Thus: "*A doubling of atmospheric CO_2 will result in a 3°C rise in global surface temperature", became the accepted scientific wisdom for use by "downstream" scientists and the public for their environmental-effect calculations.*

This is the way the scientific-acceptance system usually works. If you are, say, a teaching or research chemist or biologist, you accept the conclusions of expert climatologists on climate matters. This tradition has its roots in the words of Virgil (25 BC): "*Experto credite*" ("Believe an expert"). So even though modern science only goes back about 400 years or less, this is the historical basis for how it came to be that the scientific community endorsed the concept of anthropogenic global warming. Today, such endorsements are not unusual at all. If you were a peripheral scientist you would be doing what was expected by accepting the finding of climate experts. You might say that non-climate scientists will usually find the conclusion of the experts acceptable unless they have a compelling reason to question those findings.

In this special case, let's look in greater detail at just how the inner core of climatologists arrived at their decision and what the circumstances were that led them to be so sure of that conclusion at that particular time – and thus how this acceptance by the rest of the peripheral and general scientific communities came about.

The Unfolding Science

Perhaps the easiest way to understand this topic is to step back in time and watch the science unfold as the scientific community did, starting more than a century ago. As stated above, in those days it was known that something in the atmosphere made the earth some 33°C warmer than it would have been if the earth had no atmosphere. We call this the greenhouse effect. There are three primary gases in the earth's dry atmosphere (nitrogen at 78 percent, oxygen at 21 percent, and argon at one percent) and all have been found to have no consequential greenhouse effect. So, by default it was strongly suspected that the features causing this 33°C warming were clouds, water vapor (which varies but has a minimum concentration of about one half of one percent on up to about three percent of the earth's lower atmosphere, but is not counted as a part of the dry gases in the atmosphere), and perhaps some of the trace greenhouse gases. A prime suspect was carbon dioxide (then at just under three one hundredths of one percent and was a gas that, when looked at in isolation, had a decided greenhouse effect). Thus, any, or some combination of clouds, water vapor, and carbon dioxide might be keeping the heat of the earth from escaping. But no one really had any idea how much each of these contributed to the overall 33°C warming.

So we'll follow the simplistic path that early researchers followed during the first half of the twentieth century which was to try to determine the relative effect of water vapor versus CO_2 and apply a "cloud factor" based on the percent of cloud cover (estimated to be about 50/50 at that time) and not worry too much about clouds just yet – since they were so hard to understand and quantify.

In the late 1890s a Swedish scientist, Svante Arrhenius, wondered not only if the CO_2 that humankind was emitting into the atmosphere in ever-increasing quantities might be affecting the temperature of the earth, but also if earlier "natural" changes in atmospheric CO_2 could have affected past global "paleo" temperature excursions that caused the recurrent "glacial stages" that German geologists had recently defined based on the earth's surface geology (primarily evidence of past glaciers where none were now present). After consulting the best available minds of the day (such as Langley and Hogbom among others), he concluded that the heat-absorbing ratio in the atmosphere was 81 for water vapor to 62 for CO_2[22]. Using his conceptual data package of information and other assumptions, he developed a climate model that divided the world into fine geographic and altitude-based segments or "grid cells". He proceeded to make hand calculations for some sixteen hours a day for nearly a year before concluding that – based on the assumptions in his conceptual package – if the quantity of atmospheric CO_2 was doubled, global surface temperature would increase by about 6°C.

It was only several years later that another Swedish physicist, Anders Angstrom (as well as others), pointed out that Arrhenius's water vapor to CO_2 assessment ratio was way-off. The water vapor effect was much greater, while the CO_2 effect was much smaller than Arrhenius had assumed in his conceptual package. This cast considerable doubt on his calculations and discredited a year's arduous labor by Arrhenius.

After that, the saturating effect of water vapor became accepted by most scientists. The prevailing scientific consensus for the next half century became that the greenhouse effect was primarily caused by naturally occurring water vapor and clouds, and that CO_2 and other trace atmospheric greenhouse gases had very little, if any, influence on the earth's surface temperature[23]. In other words, the addition of carbon dioxide to the atmosphere by humans was not worth worrying about.

A Different Possible Cause

Around the time of World War I, a brilliant Serbian geophysicist and civil engineer, Miluten Milankovitch, expanding upon the earlier findings of Copernicus and Kepler, worked out the long-term changing geometric relationships between the earth and the sun. Milankovitch found four dominant long-term cycles in this earth/sun geometric relationship that influenced the subtle changes in proximity of the two bodies to each other[24]:

The longest-term cycle was eccentricity – that is, a slight change in the elliptical orbit of the earth around the sun caused by the gravitational pull on the earth of the moon and all the planets combined which very slowly pulled the earth slightly closer to the sun and then back again on about a 100,000-year cycle.

The second cycle, at about 41,000 years, was a change in tilt of the earth's axis of rotation caused by the pull of the moon alone.

The third and fourth cycles involved precession or "wobble" of the earth's axis of rotation of about 24,000 and 19,000 years and were caused by the pull of the two largest planets, Jupiter and Saturn.

Milankovitch and others wondered if the nearing and receding sun in accordance with these long-term cycles might be the cause of the glacial-to-genial-to-glacial cycles evident in the earth's surface geology that the German geologists had defined and that had puzzled Arrhenius. But the German geologists had concluded that there had been only four glacial stages in the past two million years. And this conclusion had been widely accepted by the scientific community. The closest fit for the Milankovitch cycles, as then defined (the 100,000-year cycle), would have called for twenty such stages (instead of four) in the last two million years, so this notion that the nearing sun on these long-term cycles was the cause of the glacial stages conflicted with the scientific consensus of the time and hence did not gain traction[25].

The Age of Revelation

Between the two world wars an "age of revelation" developed for scientists. Special instruments allowed humans to "see" things never seen before. X-rays were an example. One could see inside the human body or even see a flaw in a weld in steel plates. Likewise, new ingenious instruments were allowing scientists to see into the properties of materials on the atomic and molecular level – much smaller than could be seen even with an optical microscope. Molecular spectroscopy opened up a wonderland.

In 1918, the German physicist Gerhard Hettner, looking at the properties of various materials on the molecular level, saw that water vapor's spectroscopic infrared absorption band had a "dip" or "window" and noted it[26]. A decade later, the British physicist George Simpson, who was also studying that area, realized that, when applied to the earth's atmosphere, this dip or window in the water vapor heat-absorption band would allow heat from the earth to escape – and further, that CO_2 had a short band or "plug" at approximately the same wave length as the "window" in the water vapor band that could partially block some of that heat from escaping and thus perhaps atmospheric carbon dioxide could indeed have more influence on global temperature than would be deduced from just comparing the relative concentrations of the two[27]. These discoveries were a major step in setting the stage for the scientific community to begin to believe that carbon dioxide might play a consequential role in global warming. Simpson saw not only what could be happening at the molecular (micro) level but the implications at a very different (macro) level. He was knighted by the king for his perceptive discoveries.

In the early-to mid-1950s, Gilbert Plass, using a new calculating tool (a MIDAC digital computer) and formulating a more up-to-date conceptual package of facts and assumptions about the greenhouse effect, ran the global warming numbers again and it only took him a few hours to do what had taken Arrhenius nearly a year. Plass' result was a 3.6°C rise in global surface temperature for a doubling of atmospheric CO_2. But he cautioned that the molecular spectroscopy he used, particularly for water vapor, was too incomplete and preliminary for a firm result and he had only published it to demonstrate the technique, not the result[28]. As time went on, he, like many others, succumbed to the allure of the concept of human-caused global warming (without any additional validation of the original input suppositions) simply because he could see no other explanation for the global temperature changes that were occurring.

Up until this time (mid-1950s) many scientists did not worry about the possibility that human emissions of carbon dioxide might be causing global warming because although it was clear that humankind was discharging a lot of it into the atmosphere, it was not clear that this additional greenhouse gas was staying there. Many scientists thought the CO_2 might be being rolled into the ocean and dissolved or it might be being absorbed by plant life (or some of each). The prospect that it was staying in the atmosphere, but that its effect was being significantly nullified by omnipresent water vapor was apparently not being considered.

In 1957, a scientist from Scripps Institution of Oceanography, Charles D. Keeling, began to continuously record the CO_2 content of the atmosphere at the new lab on the side of the Mona Loa volcano in Hawaii. Within a decade, it became evident that carbon dioxide was indeed building up in the atmosphere. Within two decades the "Keeling curve" still showed a continuously rising atmospheric CO_2 content (see graphic 2.2). And it still shows this continuous upward trend now, after some six decades (see graphic AW.1 in the Afterword).

During the 1960s and 70s James Hansen of NASA's Goddard Institute of Space Studies (GISS) as well as Syukuro Manabe and Richard Wetherald of NOAA's Geophysical Fluid Dynamics Laboratory (GFDL) continued the search for the effect on global temperature of an atmospheric CO_2 buildup. Hettner and Simpson had shown that there could be an effect – but whether that effect was trivial or consequential was still unknown. Despite the fact that their conceptual packages were incomplete and contained unconfirmed assumptions, the lure of finding something that was of great importance to the future of the human race remained exciting. When people questioned Hansen's package of assumptions (which had resulted in a climate sensitivity figure of a 4°C global surface temperature rise for a doubling of atmospheric CO_2), for instance, he repeatedly turned to the "*but even if we're not sure, we've got to take action to stop the use of fossil fuels just in case it might be true*" argument – which apparently was to him a completely valid position. But it signaled the lack of confidence he may have had in the certainty of his figures (and in his conclusion that there was a consequential global temperature influence as a result of rising atmospheric carbon dioxide). This opened his scientific conclusion to questioning by those interested in solid credible scientific facts (and by those who had a greater appreciation of human dependence on fossil fuels and the difficulty associated with attempting to convert to something else – or even the possibility of doing so).

As stated, Hansen was getting a 4°C rise in surface temperature for a doubling of atmospheric CO_2 while Manabe and Wetherald were getting a 2°C rise with their different conceptual input data-package. These results, while quite different from each other, piqued the curiosity of many other scientists because they both fell into the "consequential" rather than "trivial" category. Still, there was considerable disagreement within the scientific community about the importance of these preliminary findings. For example, the infrared absorption band of water vapor was very preliminary, making this approach to determining the effect of atmospheric CO_2 concentration on global temperature still extremely questionable (as Plass had observed).

By 1966 Roger Revelle, who was at that time the head of Scripps Oceanographic Institution (and who later moved to Harvard University), said this notion that humankind's addition of CO_2 to the atmosphere (and its resultant impact on climate) "*should probably contain more curiosity than apprehension*"[29], while Hansen very decidedly viewed the possible results of the concept with extreme trepidation. In fact, the more resistance he got to his alarm, the more alarmed he became. His thinking seemed to be: "*Why couldn't people understand that if this is true, humankind is fouling its own nest by continuing to use fossil fuels? And we can't afford to wait the many decades it will take to see if this is true, we've got to stop using fossil fuels now*". Thus, it was easy to either jump to a dire conclusion, as Hansen had, or to dismiss this notion as too preliminary with too many unknowns and too many suppositions for concern, as Revelle had. And people in the loop were doing both with abandon.

In 1970, a 22-year-old senior at Harvard, Al Gore, who was taking a course in general environmental science from Roger Revelle, was significantly impressed when he heard about this possibility for environmental calamity (and he later became very alarmed)[30].

The Fundamental Question

In 1979, some members of the JASON group of prominent scientists who act as advisors to the government, approached President Jimmy Carter's science advisor, Frank Press (who was a highly qualified geophysicist) about the difficult-to-evaluate possibility that the increasing quantity of CO_2 in the atmosphere as a result of humankind's growing use of fossil fuels might have been causing global temperature to rise. President Carter, an engineering graduate of the United States Naval Academy and someone who was schooled in nuclear technology, listened intently and understood their concern. He appreciated the difficulty of evaluating the validity of this premise, and thought it a good question for the scientific community to research and ultimately answer. He then tasked the National Science Foundation to answer a fundamental question:

"Is the human use of fossil fuels causing the earth to warm?"

To decide how to answer this question, a committee chaired by Jule Charney, chairman of the climatology department at M.I.T, was formed and joined by eight of the world's most prestigious climatologists. With their broad knowledge of the topic they quickly homed in on and became particularly interested in four applicable pieces of information:

- The instrument temperature record of global surface temperature versus time since 1880.

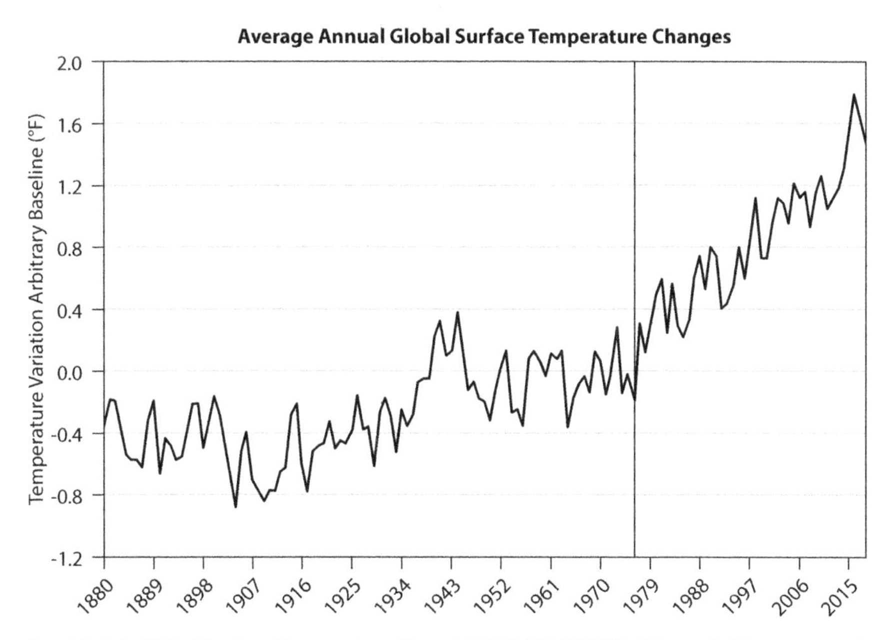

Graphic 2.1: **Global Surface Temperature Since 1880** (NASA/GISS)[31]. [Time →]. (Charney panel information stopped at the vertical line).

In graphic 2.1. Note that even in the late 1970s, global temperature shows a modest increase from the early 1900s – even though there is a slight decline from the mid-1940s to the late-1970s.

- Charles Keeling's graph of increasing atmospheric CO_2 content between the late-1950s and the 1970s (graphic 2.2).

Historical Graph: "Keeling Curve" of Atmospheric CO_2 Content.

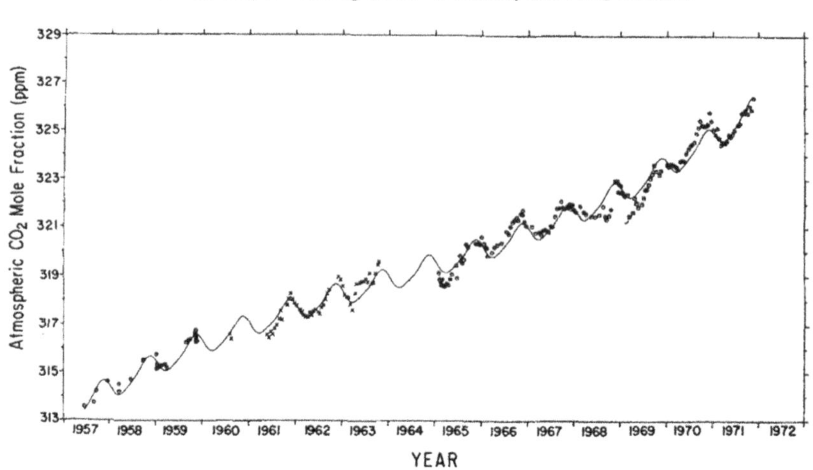

Graphic 2.2: **Late 1950s through Early 1970s "Keeling Curve" of Atmospheric CO_2 versus Time.** (C.D. Keeling/Scripps Institution of Oceanography)[32]. [Time →]

In graphic 2.2 note that the information shows a steady continuous rise in atmospheric CO_2 content, along with seasonal variability. (For the complete 1957-2020 Keeling Curve see graphic AW.1 in the Afterword).

- Manabe and Wetherald's 1970s computer climate model that predicted a 2°C rise in global surface temperature for a doubling of atmospheric CO_2.
- Hansen's 1970s computer model showing a 4°C rise for a doubling of atmospheric CO_2.

After a week of deliberations at Woods Hole, MA., the Charney Panel concluded[33]:

- The earth's atmospheric content of CO_2 is definitely rising as a result of human activities.
- The overall long-term global surface temperature is apparently rising (even though it had been falling for the last thirty-five years).
- It would be a "*rational supposition*" that if the globe is warming it could be caused by the increasing atmospheric CO_2 and if that is the case, the result could be an increase of 3 ±1.5°C for a doubling of atmospheric CO_2. (Charney later confided that this figure was arrived at by simply averaging Hansen's 4°C with Manabe and Wetherald's 2°C results – and making the error bar wide enough to include both.)
- These findings are of low certainty and a thorough scientific study and analysis should be conducted to find a more definitive answer.

The National Science Foundation, with President Carter's blessing, made an appeal to Congress to fund the requisite research. The Congress, agreeing it was an important question that potentially affected the very future of humankind, willingly agreed to do so.[34]

The Scientific Community Climbs on Board

The search for the cause of global warming was the most significant investigation the scientific community had been asked by the government to answer since the imperatives of World War II and the then-current needs of the Cold War – yet this climate search did not involve classified or secret information. Thus, it was open to almost any scientist who chose to get involved. Furthermore, it could lead to the possible saving of the human race from its own folly. And never had a creative scientific community had such a funding bonanza – an open checkbook for research on any topic that might have any bearing on global warming. Actually, much of the proposed research was important (at least to the specialist involved) and previously had encountered difficulty financially justifying itself. But now, if it could be connected to global warming, it was funded. It gave a new spark and direction to the field of science. There seemed to be an added dimension to this search – something almost spiritual.

Once it seemed evident that global surface temperature was rising on a sustained basis (despite the ups and downs that were found on a short-term basis) and that atmospheric carbon dioxide also was consistently rising as a result of humankind's carbon dioxide emissions, most of the general scientific community began to believe that the rising CO_2 had to be causing the rising global surface temperature simply because there was no other plausible explanation for why the earth was warming.

Of the other potential causes mentioned earlier, the variability of solar radiation was of a low level of scientific understanding even to solar scientists – but it looked as though changes in solar irradiance were not powerful enough to cause the changes in global surface temperature we were witnessing. In fact, bar charts showing that greenhouse gases had a considerable climate forcing effect while solar forcing was insignificant were published in prestigious peer-reviewed scientific journals[35] (see graphic 1.3). Thus, the second possible cause of global warming – increasing solar irradiance – was not considered strong enough to be the culprit.

The third potential cause – a change in the earth's albedo (reflection of incoming solar heat) caused by a subtle but sustained change in overall cloud cover had not yet been characterized or defined well enough to be scientifically evaluated. So this potential cause was not even on the climate scientists radar screen. They simply stated that clouds also were "*of a low level of scientific understanding*".

The concept of global warming being caused by changing circulatory ocean currents was vaguely conceptualized but thought to be short term and only regional.

Here we begin to see the results of a scientific philosophy called "logical positivism" (which espouses the belief that only "that which is observable" is meaningful) that had entered the scientific community early in the twentieth century. This practice which was intended to exclude "speculation" from "science" also had an unfortunate side effect – an "if it hasn't been well characterized and defined or quantified, then just ignore it" attitude rather than recognizing a lack-of-scientific-understanding as an opportunity for exploration and discovery. In other words, it put the blinders onto scientists and could be used to exclude anything from consideration that wasn't well understood and quantified (perhaps including something one didn't even want to think about because it might be beyond one's range of comprehension or might deviate from one's favored hypothesis). For these reasons, logical positivism was discarded by scientific philosophers by the mid-twentieth century. However, it seemed very much alive in the emerging science of climate change – and for all the wrong reasons.

By default, the remnants of the still extant philosophy of logical positivism left the human use of fossil fuels and the resultant increase in atmospheric greenhouse gases as the only viable candidate for the global warming we were witnessing. Such esoteric subjects as what might have caused sustained changes in cloud-cover and the many not so obvious aspects of the sun's role which were not well defined had simply been excluded before they were even considered. Virtually none of these possible causes were considered by climate scientists at that time with the exception of changes in solar irradiance which were termed "too weak to be a factor" in the changing global surface temperature we were witnessing.

Thus, changes in the sun – other than the minor changes in solar radiated heat (which were considered "too weak" to be of any consequence) – were left completely out of the discussion of possible causes of global warming and climate change. The changing quantity of greenhouse gases in the atmosphere continued to be the only applicable factor considered by the atmospherically-oriented anthropogenic climate science community of experts.

Early Study of the Sun

Humans began to systematically study the sun in the 1600s with the invention of the telescope. Sunspots and other features were observable even before early telescopes (appropriate shielding through smoke or smoked glass was highly recommended) and there is a fairly accurate record or count of these features since the early 1600s (see graphic 2.3). Sunspots, which in-and-of-themselves are cooling spots, are accompanied by other more powerful features of solar surface activity such as coronal loops, solar flares, filaments, prominences, and coronal mass ejections that give a net increase to solar irradiance. So when sunspot activity is high, solar heat radiation increases because of the increased solar surface activity. Humans have been fascinated enough by this activity to count and record sunspots. Thus, we have a 400-year record of solar surface activity which has been quantified by the number of sunspots. This gives modern researchers a feel for changes in total solar heat caused by changes in solar surface activity. Actual measurements of solar heat radiation which started in the twentieth century can be matched with the sun spot count and thus a hindcast back to the seventeenth century is possible. It is these kinds of data, which were compellingly contextualized in 1976 by John Eddy[36], combined with modern satellite data from the last few decades that make it look as though changes in solar heat radiation are not sufficient to account for the changes in global surface-temperature that we have seen in the last 400 years or so.

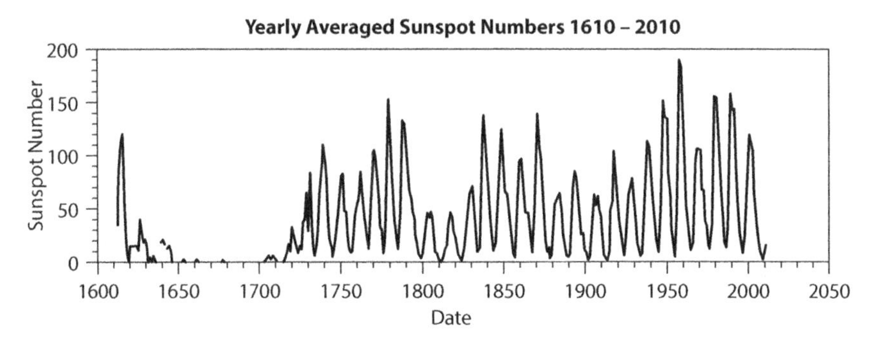

Graphic 2.3: **Yearly Sunspot Count** (NASA/MSFC)[37]. Note the extreme dip in the late 1600s. [Time →].

The solar dipolar magnetic field reversal (the 11/22 year solar magnetic reversal) and other aspects of the sun's influence on global temperature will be discussed later in chapter 15.

So if you were a scientist in the 1970s or 80s, you only had one viable option to explain rising global surface temperature and that was rising atmospheric carbon dioxide as a result of the human use of fossil fuels. Thus, in order to answer the President Carter question of humankind's role in global warming it became incumbent upon the climate science community to go find and flesh-out the evidence that pointed to human use of fossil fuels so that they could explain to the skeptics what the climate science community thought they already knew – that is, their foregone conclusion that human-caused increasing atmospheric greenhouse gases were responsible. Thus began the effort to "fit" the evidence that was available to the premise (or "*rational supposition*") that existed at that time – that an increasing amount of atmospheric CO_2 content was causing global warming. This proved to not be an easy task, as you will soon see.

The Marine Sediments

Hardly considered by atmospherically oriented climatologists, was the key geological discovery in 1976 by the marine geoscientists Hayes, Imbrie, and Shackleton, who after analyzing two Indian Ocean deep sediment cores, determined that there had been four glacial stages in the last 460,000 years – roughly in sync with the 100,000-year Milankovitch solar cycle[38]. As well, geologist George Kukla found the same thing in the stratification of clay in the open pit clay (brick) mines and riverbanks of Central Europe[39], and geology professor Wallace Broecker of Columbia University found the same thing in the reef terraces of Barbados[40] and New Guinea[41]. Thus, it became quite clear that an approximately 100,000-year solar cycle as originally defined by Milankovitch was the "pacemaker" of the ice ages and that the German surface geologists of many decades earlier (and the scientific consensus of the time) had been wrong when they postulated only four glacial stages in the last two million years. It became apparent that subsequent glacial events probably had obliterated the surface evidence of many of the earlier ones.

Meanwhile, the notion that humans and their fossil fuels were responsible for global warming was gaining momentum within the scientific community – particularly in the biological arena. More and more scientists thought they could see a connection between their favorite subject of research and the warming globe – and they enthusiastically offered-up grant proposals for research on the potential effects of such a perceived eventuality. Since most of these biological scientists still thought it had to be atmospheric greenhouse gases causing the warming, the anthropogenic notion further reinforced itself, even though what they were seeing was a warming earth that could just as likely be warming from natural causes as by human activity.

Preconception Bias

Enthusiasm over the topic of global warming spilled over into the journalistic and environmental communities as well as the general public. This led to political interest and, of course, involvement by policy-makers.

Global warming as a result of human behavior was a simple and seductive concept. The commonly held logic for a huge segment of the educated and informed public throughout the world would soon become:

- *We can see that the globe is warming. The Arctic ice is melting. Polar bears are confused and hungry. Each year there seem to be more wildfires in California.*
- *Carbon dioxide, a greenhouse gas, is pouring from our tailpipes and coal-burning power plants. The overwhelming majority of scientists tell us that the rising global temperature we can see and feel is caused by that ever-increasing atmospheric* CO_2.
- *With all this evidence, how can anyone tell us that human use of fossil fuels is not causing global warming and climate change?*

The missing information that key scientists somehow had not recognized that much of CO_2's effect would be nullified by atmospheric water vapor which was already absorbing much of the heat being radiated outwardly by the earth, was not being discussed. In addition, the influences of the sun, with its changing powerful magnetic field, and other gravitational and radiological influences were simply not generally understood to be important forcers of global temperature change and thus were not being discussed (we will get into these in more detail in Part II of this book). This was at least partially because these qualities, some of which were suspected or even somewhat known to the scientific community, were not yet "well-understood" and largely not quantified.

Ida R. Hoos, an astute twentieth century sociologist and social critic, said it perhaps better than anyone as she waved her perceptive finger at the scientific community: "*What can't be counted simply doesn't count*"[42]. Ida proved to be very observant.

So it seemed to many that increasing atmospheric CO_2 was causing rising global temperature. And the peripheral scientists and statisticians who had accepted this premise were busy giving us all sorts of quantified and often alarming downstream environmental effects. The momentum of the anthropogenic premise slowly built as more and more people fell under its spell. After all, the thread of that logic-path seemed to be validating itself, for as humankind emitted more and more CO_2 into the atmosphere, the earth seemed to be getting warmer and warmer. And many people, including many in the scientific community, who were standing on the early plateau of knowledge derived from the logic-path mentioned above, still believed and followed this easy-to-understand reasoning. This presumed cause-and-effect relationship that seemed so real to so many, led to discussions, arguments, and counter-assertions with the non-believers: These petty disputes would arise with each undulation in global surface temperature because none of these quibblers, on either side of the controversy had a good grasp of what the real causes of these temperature fluctuations were.

Judgment Science

There was another powerful factor that influenced the eventual general acceptance by the inner-core of climatologists that human beings and their emissions of carbon dioxide into the atmosphere was causing global warming: This is that "*judgmental assessment*" was often used in the close-cousin to climate science – meteorology or weather forecasting.

Many climate scientists had studied meteorology which had been a slowly developing branch of science over many years, propelled, no doubt, not only by the inconvenience of bad weather but also by the terrible tragedy of erratic and unpredicted storms.

Throughout history, as people across the world became more aware of global geography, primarily through maritime activity, news of massive and disastrous weather events slowly spread, for example:

- An unusual tropical cyclone (hurricane) made its way to London in 1703 with the loss of only 128 people on land, but the loss of 15,000 sheep and 400 windmills. In the rivers Thames and Severn, however, over 700 ships and 8000 seamen were lost because the storm had not been predicted[43].
- In 1715, the entire eleven-ship Spanish treasure fleet, which was returning from Cuba to Spain was lost to a hurricane off the Florida coast along with hundreds of millions of pesos in gold and silver bars and coins as well as precious jewels and Asiatic porcelain[44]. Also lost were over 1000 crew and passengers – all because this hurricane was unpredicted. This event almost bankrupted the Spanish throne which was counting on the shipment to pay off debt incurred in a recent war.

- In 1780 an unpredicted massive hurricane swept through the Caribbean Sea impacting island after island. When the aftermath was assessed, it became apparent that over 22,000 people on multiple islands had perished[45].

- The Bay of Bengal (bordered by India, Bangladesh, and Myanmar) was (and still is) particularly vulnerable as cyclones approach from the south into shallower water and the storm surge rises: In 1789, some 20,000 people perished as a result of such a cyclone; in 1864, it was 40,000 people; in 1867 it was 100,000 people lost as the storm surge increased from 10 feet to 50 feet; in Bangladesh in 1970, the Bhoda cyclone killed between 224,000 and 500,000 people; however, in 1991 a far more destructive cyclone (ten times the property damage of the Bhoda) killed only 138,000 people because of improved warning, protection, and evacuation plans.

The Great Indonesian Tsunami

All of these "weather related" events could be contrasted with a geophysical disaster, the "Great Indonesian Tsunami" of 2004 when 228,000 people, primarily in Aceh Province of Indonesia, were killed by a massive and unexpected undersea megathrust (earthquake) off the coast of Sumatra. It lasted for almost ten minutes and induced a 100-foot-high tsunami (wave train). This catastrophe then led to increased efforts to predict and warn of such geological events)[46].

The worst tropical storm to hit the United States was the Galveston (Texas) storm of 1900 which not only destroyed most of the city but also killed some 8000 people[47]. A near repeat of this storm in 2017, caused the death of only 50 people because the storm was appropriately predicted and the capability to evacuate the most endangered existed and was accomplished (but that did not stop the billions of dollars in property damage).

So, as you can see, there was a growing incentive to use whatever means necessary to predict these kinds of storms. In the case of the Galveston hurricane of 1900, the means to have predicted it were there (led by the telegraph). But mostly due to arrogance, the United States National Weather Service did not believe the Cuban reports that the storm was headed toward Texas because the "experience" and "judgement" of the Americans told them it would head north before that. Unfortunately, it did not, and the city of Galveston was totally unprepared for a storm of such destructive force despite the fact that the U.S. had one of the most technologically up-to-date and well-organized weather forecasting capabilities in the world at the time. They had been warned about the storm by the Cubans, but failed to warn the public because they relied too much on their own judgment and decided not to recognize the Cubans' well-considered analysis[48].

Over the centuries, anyone who could prognosticate significant weather events was given a near-reverential status. Early on, such prediction was considered a "black art", and hunches and forebodings prevailed and sometimes even proved to be true. As the science progressed from philosophical to astrological to observational, so-called expert judgement grew in stature in the field of weather forecasting.

So, as we gravitated into climate science there was a history of judgmental assessment being used in atmospheric science and that strong dependence on judgement was carried over into the new field of climate science. Unfortunately, judgement can be strongly slanted when it is based on preconceived but unsubstantiated judgements (as in the case of Galveston), and it is sometimes difficult in the climate science business to tell where the judgement science ends and the absolute science takes over. Thus, the potential existed to confuse the two and for those who were so inclined, to attribute absolute standards to judgmental conclusions. Unfortunately, this is what happened as preconceived judgements strongly influenced the explanation of the causes of global warming and climate change.

Uncertainty

By the late 1970s there was some confusion within the scientific community about global warming and what was causing it. However, among many climate scientists, there was a strong undercurrent of preconceived but unsubstantiated conviction that human use of fossil fuels emitting carbon dioxide into the atmosphere was the cause.

During the 1980s, the study of global warming (under the impetus of the JASONS/Jimmy Carter/Charney Panel/U.S. Congress initiative) was the subject of considerable scientific funding mostly by the United States government and mostly aimed at the wrong question: Is the earth warming? Rather than: If the earth is warming, what is causing that warming? No one was overseeing the overall direction of the government's inquiry except to make sure the research was somehow related to global warming or climate change. So, in addition to the legitimate scientific inquiry, a number of scientists scrambled to see if their favorite research topic could be considered to have a bearing on "global warming" (or as this phenomenon was increasingly being called by environmentalists, "climate change") so they could receive grant support from one of the thirteen U.S. government agencies that were doling out this largesse. And with almost unlimited funds available, in perhaps too many instances they were successful.

Chapter 3

The Mounting Case for Human-Caused Global Warming

> *A small body of determined spirits fired by an unquenchable faith in their mission can alter the course of history.*
>
> — Mohandas Gandhi, 1946

By 1988 hundreds of diverse but at least remotely applicable climate research papers were being published. Most were proxy studies to show that the globe was indeed warming – such as a study showing that arctic birds were moving their nesting-range north, presumably because the old nesting-range had become too warm. Or another paper showing that butterflies were moving their range north in England (or another – up the mountainsides in Scotland) to get away from the heat. Or, more tragically, the account of the montane golden toads of Costa Rica, who apparently were unable to adapt or migrate, and simply went extinct[49]. And there were hundreds and hundreds more fascinating scientific papers about global warming. It seemed perfectly natural for biologists to attribute this climate change to human-use of fossil fuels because that was the more-or-less established conventional wisdom. However, many of them seemed to conclude that the types of biological response that they were so meticulously studying somehow verified that increasing atmospheric greenhouse gases were causing global warming, without stopping to think that there might be other possible causes for that warming. These studies were simply proxies verifying what we already knew from the instrument record – the globe was warming, and noting the biological response to that warming – but not investigating anything about *what was actually causing* the earth to warm. Perhaps 85 percent of all the research on global warming being sponsored by the United States government at that early time was simply trying to proxy verify that the earth was indeed warming – and often noted the various, sometimes tragic, biological responses to that warming.

The IPCC (Intergovernmental Panel on Climate Change) and The UNFCCC (United Nations Framework Convention on Climate Change)

What was needed was an impartial body to take a balanced look and act as a neutral observer. So it was decided by all concerned to form a group to assess all the global warming information, to summarize findings, and to make reports. Since global warming was a worldwide problem it was decided to make this group an international panel. Thus, the Intergovernmental Panel on Climate Change (IPCC) was formed by the World Meteorological Organization and the United Nations in December 1988. From its inception, however, this organization proved to be far different than a neutral sorter of information on the topic of global warming. The UN Framework Convention on Climate Change (UNFCCC) intergovernmental treaty that was drafted in 1992 was based on the First Assessment Report of the IPCC in 1990. Its charter illustrates the depth of the widespread preconception and bias on this topic within the international climate-science community at that time[50]:

- First, climate change was defined as "*a change in climate caused by the activities of man.*" (While, of course, the human role in creating any effect on climate had not been scientifically established, although it was a sort of presumed supposition among the inner-core of climate scientists).
- Second, the ultimate objective of the UNFCCC was stated to be to "*stabilize greenhouse gas concentrations in the atmosphere at a level that would prevent dangerous anthropogenic (i.e. human induced) interference with the climate system.*" Again, this objective was stated before it had been established that changes in atmospheric CO_2 or any other greenhouse gas content in the atmosphere had any effect on global temperature or the climate. Later, it was this thinking that led to the interpretation within the IPCC that no cause of global warming other than human-caused emissions of greenhouse gases could be considered or discussed in IPCC reports.

These definitions showed a bias based on a preconception on the part of the founders of the IPCC and the UNFCCC – nearly all of whom had already subscribed to the notion that humans (and the CO_2 from their fossil fuels) were responsible for global warming (despite the lack of any scientific basis for this premise). This foregone conclusion had become established primarily because humans were pouring huge quantities of carbon dioxide into the air, global temperature was rising, and therefore one must be causing the other. This thinking prompted one respected researcher to state: "*The conclusion has been drawn before the research has begun.*"

Climate Change? Global Warming?

The terms "climate change" and "global warming" are perhaps the most misunderstood definitions in the global warming lexicon. When we say "climate change" do we mean a change in climate as a result of humankind's activities (which is the UNFCCC's definition) or do we simply mean a change in the earth's climate from any cause including natural causes such as increasing solar irradiance or changing cloudiness and the resulting reflectivity of solar heat radiation? The UNFCCC's definition seems to imply that there is no other possible cause than human activities causing increasing atmospheric greenhouse gases thus cementing the notion of human use of fossil fuels as the culprit. The same holds true for many people who seem to think that "global warming" means "global warming as the result of human-caused emission of greenhouse gases into the atmosphere".

Soon the IPCC – which was becoming a huge, many-faceted, multi-national loose- aggregation of various scientists, environmentalists, economists, public-awareness (including "public-opinion guidance") experts, and political advocates – had three distinct functions:

- The public-opinion guidance function to publicize the generally presumed notion (among the inner core of climate scientists) that rising atmospheric CO_2 caused global warming, and to warn humankind that they were polluting the atmosphere with harmful CO_2 by using fossil fuels and causing global warming (now frequently called: climate change).

- The scientific function to review scientific papers that examined the environmental and human effects of a predictably rising global temperature – presumably caused by humankind and its predictable emissions of greenhouse gases (such as CO_2) into the atmosphere – and to produce various assessment reports.

- To monitor and evaluate research on whether or not human-caused rising atmospheric CO_2 caused global warming.

Hence, this was a scientific version of "putting the cart before the horse". Of these three functions, most of the funding was decidedly on the first two. Thus, the scientific community had successfully "turned" the objective that the President and the U.S. Congress thought they were funding (to determine if humans were causing global warming), into the funding of investigations that suited the scientists' own inclination and interests, but could be associated (at least somehow) with global warming.

The First Assessment Report (FAR) by the IPCC in 1990, while basically accepting that humankind's emission of greenhouse gases was causing consequential global warming, also acknowledged the need to scientifically demonstrate that these emissions were causing the globe to warm. *They stated that it would take about a decade to confirm this premise.*[51]

At that point we had a very unfortunate disconnect: (a) we still needed to demonstrate that the human use of fossil fuels and the resulting atmospheric CO_2 was a consequential cause of global warming, and (b) the dominant mission of the group that had been charged with the responsibility to make this determination was to advocate that humans stop using fossil fuels in order to lower atmospheric CO_2, which they already had assumed was the cause of global warming, as well as to research the environmental and human effects of this global warming that was presumed to have an anthropogenic cause.

Much of the $2 billion per year spent by the U.S. government on the research of climate change during those years (1990s), was going into supercomputers and their associated programming and ancillary costs with the hope that the complex causes of climate could be conceptualized, mathematically defined, and calculated. But much of it was also going into hundreds of studies by scores of scientists most of whom were simply attempting to show that the earth was warming by investigating proxies without trying to determine the true cause of that warming (since in most instances they had already subscribed to the supposition that a human- caused output of CO_2 was causing the globe to warm) – or were researching the effects of the warming globe on the environment or on people. This included everything from the change of volume of meltwater in the glacial lakes of the Himalayas[52] to the projected displacement of people in Bangladesh[53] as a result of the predicted impending rise in sea level. If you put the words "global warming" or "climate change" into a grant proposal, at any of the thirteen U.S. agencies concerned with this projected phenomenon, it was nearly always funded.

It was during this period (1980s and 90s) that the United States became a leading part of the global research engine for the concept of anthropogenic global warming. Many projects were located outside the United States with the lead scientist from one country and the support scientists from other countries (usually with one or more U.S. scientists mixed in). Often, most (if not all) of the costs were being picked up by the generous anthropogenic-leaning financial leader, the U.S. government. The socially accelerating concept of human-caused warming (the "anthropogenic global warming movement") was being promulgated, institutionalized, and internationalized by a consortium of scientifically powerful countries. In addition to the United States, scientists from Great Britain were becoming a driving intellectual force. Scientists from France and Germany also were strong participants. Many of the inner core of anthropogenic climate scientists both in the United States and internationally seemed to be becoming "possessed" by an enrapturing belief in their cause – which began to make the movement more of a moral crusade than an impartial scientific investigation. Scientific paper after scientific paper explained the prognosticated adverse effects of a predictably warming world caused by human activity – yet few, if any, attempted to demonstrate how it was known that this global warming was caused by human activity. That part was simply an assumption that virtually everyone seemed to accept as a valid starting point for their research.

As the anthropogenic premise gained traction, this wide variety of peripheral research added to the base of the pyramid of scientific knowledge that applied to the subject, but very little of the research directly applied to answering the question that kicked it all off:

Is the human use of fossil fuels causing the earth to warm?

The entire scientific foundation for the notion of human-caused global warming and resultant climate change rests on the premise, or supposition, *that rising atmospheric carbon dioxide content causes consequential rising global surface temperature.* There is, of course, no doubt that carbon dioxide, in isolation, is a consequential greenhouse gas. The salient question here is: when such a tiny amount of it is in the earth's atmosphere, where there are other *overlapping* molecular heat absorption band greenhouse gases (such as water vapor), is it still a consequential greenhouse gas, or does the water vapor absorb much of the outgoing heat from the earth that might otherwise be absorbed by carbon dioxide if there were no water vapor present? Key scientists accepted the supposition that it was just as powerful as it was in isolation (except its effect had to be split in half where it overlapped water vapor's infrared absorption band), as a valid one for the computer models they developed that calculated the effect on global temperature of varying quantities of atmospheric carbon dioxide[54]. And the results of these models became the basis for their alarm about the possibility that increasing use of fossil fuels would cause the earth to consequentially warm. But they must have known that this premise contained unestablished suppositions that might or might not be valid. This is because they sometimes stressed that even if one or more of these assumptions later proved not to be true, just the *potential* for environmental damage to the earth was so severe that we should stop using fossil fuels right-away (in case their unestablished suppositions might later turn out to be true[55]).

But the only way to properly answer President Jimmy Carter's question about human-caused greenhouse gases causing global warming, was to scientifically verify (or show to be incorrect) the premise or assumption that rising atmospheric CO_2 was a consequential cause of the rising global temperature that we were seeing. It's easy to accept this supposition as a possibility, but it is far more difficult to scientifically confirm its validity. During this time, changes caused by the sun were simply not considered by these atmosphere-oriented climate scientists to be powerful enough to cause the global warming we were witnessing on earth. This was because they were only considering changes in solar heat irradiance, not the full range of solar magnetic, radiational, and proximity influences on global temperature which were not well understood at that time. We will examine these later in Part II of this book.

Are Human Beings Responsible for Global Warming?

Nearly all scientists agree that there are three types of scientific evidence that could be used to substantiate the premise that the human emission of CO_2 into the atmosphere caused by using fossil fuels is the principal cause of a warming earth:

- Theoretical evidence (the physics).
- Observational evidence that confirms the theoretical evidence.
- Computer climate models as a substitute for full-scale experimentation (or observational evidence if none could be found).

The only *theoretical* way to determine the effect of CO_2 on the earth's surface temperature is to determine how much of the greenhouse effect can be attributed to atmospheric CO_2 (as opposed to the two primary potentially competing causes - clouds and water vapor). To make this determination it is necessary to make a number of assumptions:

The first assumption is an estimate of how much of the greenhouse effect do we attribute to clouds and how much to clear-sky greenhouse gases? Clouds are difficult to understand and quantify. Measurements of cloud cover by satellite (only available since 1979) show a cloudiness of about 62 percent (now)[56], although earlier eyeball counts showed about 50 percent and later 60 percent. But clouds have varying degrees of permeability by various parts of the electromagnetic spectrum so what you see is not necessarily what the narrow part of the affected infrared spectrum sees. Respected climate scientists Kiehl and Trenberth, in their landmark 1997 paper[57] (discussed later), showed 62 percent cloud cover but only 21.5 percent of incoming solar irradiance reflected by clouds. And these authors not only hedged on the accuracy of their numbers, but they emphasized the uncertainties. Other researchers used considerably different figures. Thus, the effect of clouds was not easy to fully understand and quantify, and there was some doubt over just how much clouds influenced the greenhouse effect.

To partition the remaining portion of the atmosphere (the clear-sky portion) between its presumed two major greenhouse-effect contributors (water vapor and CO_2), researchers had to determine how much of the earth's outgoing heat was absorbed by each. To make this determination it was necessary to make a number of assumptions and judgements that led different researchers to different answers to the amount of global warming that would be experienced if atmospheric CO_2 was doubled (ranging all the way from a 0.1°C to about an 8°C rise in global surface temperature.)[58] Thus the answers ranged all the way from the "trivial" to the "consequential" to the "outrageous". Accordingly, there was again a great deal of uncertainty. Recognizing this, many climate scientists concluded that this theoretical calculation would not give a definitive enough answer to the important question: *is human use of fossil fuels causing the earth to warm?* If, as the-end-of-the-twentieth-century deadline for an answer to the question neared, the best answer the scientific community could come up with after spending nearly $50 billion over several decades was: "*Well, yes, probably somewhere between 0.1 and 8°C*", then they would have completely failed in their task to answer Jimmy Carter's question. And the climate-science community knew they would take a loss of credibility from which they might never recover.

These descriptive categories (trivial, consequential, and outrageous) are rather vague and have not been quantified because their limits are not strictly defined by any official body, but for general-thinking purposes we could define them thus: *trivial* is so little that the environmental impact on the world would be just that: *trivial* – and that would mean a climate sensitivity figure from about just over zero (0.1°C) up to about 1.5°C or so (rise in global surface temperature for a doubling of atmospheric CO_2); *consequential* would be from about 1.5 to 2°C or so – up to about 4 or 5°C or so; and *outrageous* would be anything more than that.

Mixed into all this had to be the thinking of Miskolczi who made a scientifically plausible case that it is actually *zero* – that changing atmospheric CO_2 content has no effect whatsoever on global temperature (see textbox on Miskolczi the dissenter.).

Miskolczi the Dissenter

Miskolczi assumes that climate system changes involve constant thermodynamic equilibrium and changing relative humidity, while other climate scientists assume constant relative humidity and changing thermodynamic equilibrium. Neither of these alternatives has been established as the definitive mechanism. If Miskolczi is correct, the whole concept of anthropogenic global warming dies an instant death, since changing quantities of atmospheric carbon dioxide would have no effect on global surface temperature[59].

In any event, the truth, as it best could be determined from the physics at that time (anywhere from zero to 8°C or more) was not an acceptable answer, since it didn't answer the question about humankind's role in global warming at all.

Accordingly, as the end of the decade of the 1990s neared, it became more and more apparent to the inner core of climate scientists that they had to settle on a more firm-and-fixed figure for the theoretical answer, and they had to demonstrate observational evidence and computer model corroboration that would confirm that figure. After all, many tens of billions of taxpayer dollars had been spent. Yet of the tremendous amount of "research" conducted in the name of answering that question, nearly all of it had been expended on three categories:

- Exploring proxy methods of showing the earth was warming.
- Revealing the presumed adverse environmental and human effects of anthropogenic global warming.
- Procurement of cutting-edge supercomputers to assist in the above.

But very little of the research applied directly to the question of the human role in global warming by the use of fossil fuels (except some computer climate model analogues, which showed results that were dependent on postulated but unconfirmed input assumptions).

Thus, climate scientists, although they felt they had gone more deeply into the subject, were not really much closer to an answer than they had been two decades earlier when the research (and the tremendous expenditures) began. And in the intervening years the few researchers actually studying the question of the greenhouse gas role in global warming were still coming up with climate sensitivity results of from about zero to about an 8°C rise in global temperature for a doubling of atmospheric CO_2 depending on which set of assumptions they used in their calculations. And these results still ranged from the trivial (inconsequential) to the consequential (significant) to the outrageous (unbelievable). The inner core of climate experts knew that they needed an answer that was a consequential amount (around 3°C) to conform to what the public awareness section of the IPCC had been telling everyone for years – that the human addition of CO_2 to the atmosphere was causing a consequential global surface temperature increase. Yet they had no physics basis for making such a claim.

At that point (the late 1990s) the notion of human responsibility for the global warming we were witnessing could indeed best be called a "foregone conclusion" – because it was something that nearly everyone accepted without any scientifically demonstrated basis other than the notion that nearly all scientists seemed to believe it. Remember, there was no apparent alternative reason for global warming other than the possibility of human-generated rising atmospheric greenhouse gases because of the limited extent of understanding in the scientific community at that time of the other potential causes (such as changing solar magnetic effects which influenced global cloud cover which changed the amount of solar heat reflected away from the earth). It is probably true that most scientists went along with the premise that humankind was causing global warming, but that was because that belief had become the conventional wisdom and they could see no other alternative.

Clearly this "decision by exclusion" was not an acceptable scientific explanation after so much expenditure and effort by the scientific community to find a science-based answer. Unfortunately, most of the "science" had been directed at the wrong questions: (a) Was the earth warming? and, (b) What were the environmental consequences of that warming? The first question had already been answered by the instrument record (yes, the globe is warming), and nearly all this proxy research was simply confirming that the globe was warming.

The second category of costly research (consequences of the warming) involved projecting changing atmospheric CO_2 levels and the resulting global surface temperature changes. Then the result of that change could be assessed and evaluated for a variety of qualities vital to the human race – such as a rise in the level of the sea (and its accompanying human displacement) or the highly disputed findings of: more frequent hurricanes, more rain, massive flooding (sometimes offset by increased drought and raging wildfires) – and on it goes. Temperature extremes were now being blamed on humankind's use of fossil fuels and the resulting emission of "harmful" carbon dioxide into the atmosphere. Although there was often a scientific basis for these events being caused by a warming globe, *there was still no valid scientific basis for concluding that the warming globe was the result of human-caused increasing greenhouse gases in the atmosphere.*

The other big expenditure had been on the new cutting-edge supercomputers (and their associated software) that at first had looked so promising, until the slow realization that even they were no match for the complexities of world climate and to make them so would require so much computer power that we might contribute to global warming just to make the calculations[60].

Before the publication of the "definitive" Kiehl and Trenberth 1997 paper[61], the theoretical evidence was on very unsure ground with many different results as we mentioned earlier. Much of the computer modeling of climate radiative forcing showed a high degree of uncertainty which led to very uncertain projections of the climate-sensitivity figure. Perhaps the observational evidence could be used by the inner-core to try to convince themselves, as well as the rest of the scientific community (and the public) – that humankind's emission of greenhouse gases was responsible for global warming. So, the inner core of climatologists began to switch their public emphasis away from the theoretical evidence to the observational evidence.

The Observational Evidence

At that time, (late 1990s), there were two major talking-point pieces of observational evidence of anthropogenic global warming. One was the recently introduced flat-handled hockey stick that showed atmospheric global temperature more-or-less steady for the last 1000 years, and then it began to sharply rise about a century ago – more-or-less matching the rise in atmospheric CO_2 that was caused by humankind's use of fossil fuels. (See graphic 3.1)

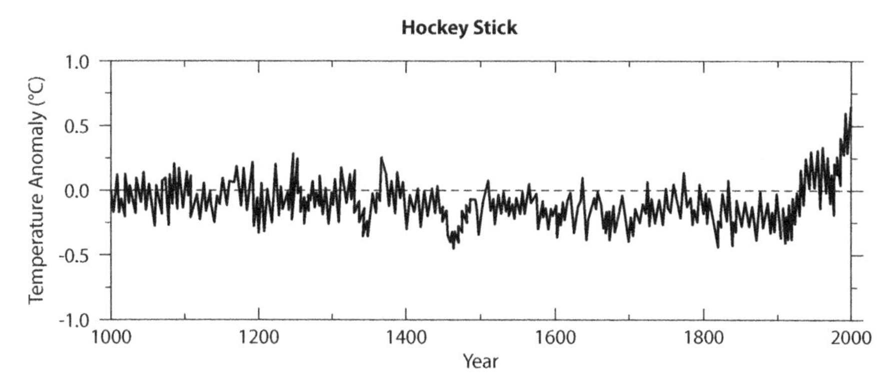

Graphic 3.1. **The 2001 - IPCC Temperature Hockey Stick** (Adapted from IPCC Third Assessment Report[62]. [Time →]

A second prime piece of observational evidence also began to be seriously discussed: In the mid-1990s the first significant data from the long-awaited *Vostok deep ice-core*, (the Russian/French/U.S. – Antarctic ice-coring project) became available[63]. Those data further confirmed the results of the Hayes, Imbrie, and Shackleton marine sediment-cores, the Kukla clay-mine findings, and the Broecker reef-terrace findings: the 100,000-year Milankovitch solar cycle was the pacemaker of the ice ages. But these ice-core data contained something that the other geologic data did not. They had matching atmospheric CO_2 content (obtained from the air bubbles in the ice) versus time, along with global surface temperature versus time – and the two (i.e. the temperature and the atmospheric CO_2 content) seemed to rise and fall in sync over each 100,000-year cycle – thus implying a cause and effect relationship. Most people simply assumed that the rising and falling atmospheric CO_2 was driving the rising and falling atmospheric temperature. (See graphic 3.2)

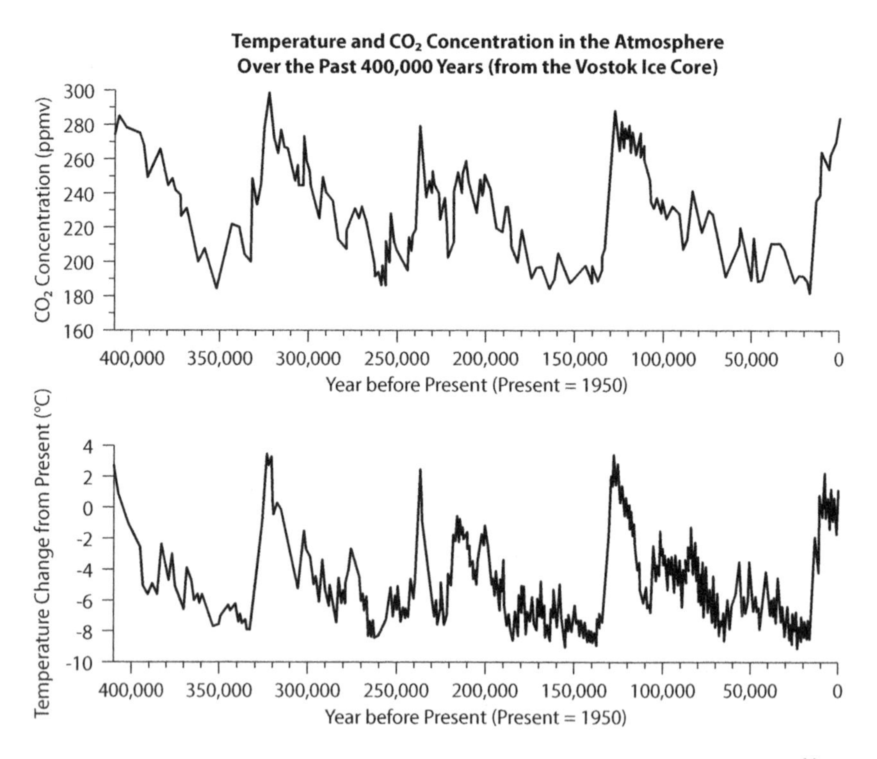

Graphic 3.2: **Vostok Ice-Core Data.** J.R. Petit/Nature; P. Rekacewicz /UNEP/GRID-Arendal[64]. [Time →].

Observationally, things were really looking good for the anthropogenic inner-core. Not only did they have a graph showing that the earth's temperature followed the atmospheric CO_2 content up and down between genial and glacial stages for the last 400,000 years, but they also had the hockey stick graphic that showed that the earth's temperature was seemingly influenced by the atmospheric CO_2 content for the last 1000 years.

Computer Climate Models

The computer climate model had become a focus of research in this field with one research group after another justifying a new supercomputer until by the mid-1990s there were over twenty in use – many using essentially the same climate model (the "community" model). Not surprisingly, with the foregone conclusion that humankind was responsible for global warming thoroughly imbued in the minds of modelers by the constant *human-caused* global warming reports from the IPCC public-awareness section, virtually all of these computer climate models showed that anthropogenic atmospheric CO_2 caused global warming. So at this point there was good observational evidence and apparently good computer climate model evidence that humankind was responsible for global warming. The only thing lacking was the basic theoretical verifier – some form of evidence that would provide a sound physics basis and a quantified answer that they could all agree to within the bounds of a consequential effect.

The Missing Clincher

Prospects of impending success surged as the *theoretical* evidence unexpectedly became resurrected when Kiehl and Trenberth, in 1997, wrote their key paper[65] (mentioned earlier) on the earth's energy budget. This paper provided a possible method to partition and quantify the open-sky (cloud-free portion) greenhouse effect. It suggested that recently observed changes in atmospheric CO_2 could cause global temperature changes that fell into the "consequential" category and thus the conclusion could be drawn that the principal cause of global warming was human-caused atmospheric gases. It was eaten-up by the inner-core like manna from heaven. Kiehl and Trenberth not only gave what looked like an acceptable figure for cloud-cover but also suggested that the open-sky portion of the greenhouse effect was 60 percent caused by water vapor; 26 percent caused by CO_2; and 14 percent caused by other trace greenhouse gases. After some additional simple arithmetic (by others) this resulted in a global surface temperature rise of 1°C for a doubling of atmospheric CO_2. When further supplemented by a calculated water vapor feedback amplification factor (caused by evaporation from the world's warming oceans and moist land areas) this added an additional 2°C increase in global temperature and gave a total temperature rise of 3°C. This quantification of the effect provided a very useful scientific tool. From this, it would be possible to use the finding in ancillary research that required quantification, and it would allow a suggestive answer to Jimmy Carter's original question:

Is human use of fossil fuels causing the earth to warm?

Here was a theoretical answer that the team could run with. Not only was it consequential and quantified, it was (coincidentally) the same number (3°C) that the Charney Panel had come up with eighteen years earlier by simply averaging two off-the-wall preliminary speculative numbers[66]. And the inner-core gave this new 3°C answer an acceptable level of certainty by stating that it was "easy to calculate and uncontested"[67]. And for the broader secondary and downstream scientific community here was an answer that was quantified and could be used for all sorts of peripheral research such as projections of future sea level rise based on prognostications of humankind's anticipated use-rate of fossil fuels and the resulting extrapolated atmospheric content of carbon dioxide.

As the turn of the twenty-first century (and the deadline for an answer to the original Jimmy Carter question) approached, the inner-core of climatologists had reached a pivotal point in the quest to determine whether human use of fossil fuels was causing global warming.

In a world of possibilities and perhaps even probabilities, when something is accepted by most of the few experts in the field it becomes "generally-accepted". And as a rule, the rest of the scientific community customarily honors it ("*experto credite*" – as mentioned earlier). In this case the inner core had been pretty much convinced from the very beginning that there was a consequential increase in global surface temperature as a result of increasing CO_2 because there was simply no other explanation (known to them) for the current warming of global temperature. Thus, their task became to gather enough supporting evidence to convince themselves – as well as the rest of the scientific community – of that. Now that point had been reached (however tenuously). The available suggestive evidence from theoretical, observational, and computer climate models had pretty much peaked and all three pointed to the human-caused position:

- The inner core had the Kiehl and Trenberth 1997 paper which they felt would cover the basic physics[68].
- They also had what, at that time, appeared to be acceptable observational evidence in the newly revised and introduced hockey stick graph with a steady global surface temperature

curve going back 1000 years and then a quick rise that pretty much matched the time of humankind's atmospheric emissions of CO_2 (graphic 3.1)[69]; and also in the Vostok ice-core curves going back 420,000 years that showed matching cycles or "excursions" of atmospheric CO_2 content and global temperature on about a 100,000 year cycle (see graphic 3.2)[70].

- In addition, they had all sorts of confirming computer climate-model evidence (which was becoming an alternate world unto itself).

Although none of these forms of evidence were by themselves definitive (all of them, at best, were based on low-level-of-certainty "suggestive" evidence) – the combination, however, was remarkably seductive. Phrases such as "many different forms of evidence", and "multiple lines of investigation" were being widely used by the media. So multiple "almost-good-enough" lines of evidence were bundled-up to be "good-enough" if looked at collectively. And the IPCC's self-imposed deadline for an answer to Jimmy Carter's question was rapidly approaching. Furthermore, based on the unrelenting IPCC "awareness" barrage of some twelve years (and the fact that there was no known alternative to the anthropogenic cause of global warming), most of the peripheral and general scientific communities already believed it was an established fact that humans were responsible for global warming. Now even the mildly skeptical scientific community, along with the journalistic community, and the general public were becoming swayed. Some knowledgeable neutralists were awaiting final confirmation by the inner core of climate experts.

Since the collective evidence and the scientific consensus were probably as good as they were going to get and the inner core's own deadline was upon them, the time for them to formally proclaim that humans were responsible for global warming seemed to be at hand.

Meanwhile, however, the IPCC summarizers of the Detection and Attribution (D&A) section who were simply reviewing the technical data that was available from research papers to draw their conclusions, had gone on their merry way and answered three fundamental questions[71] for the about-to-be-released IPCC Third Assessment Report that was due out at the turn of the century:

- We are 100 percent sure that human use of fossil fuels is causing atmospheric CO_2 content to increase.
- We are about 80 percent or so sure that Global Surface Temperature is increasing.
- We are about 66 percent or so sure that humankind's fossil fuels are causing the globe to warm.

Since we weren't there, we can only imagine the reaction when key members of the anthropogenic global warming inner core of climate scientists reviewed the as-yet unreleased but finalized and ready-to-go report:

> *"They've got to be kidding, this is madness..." "We're only 66 percent or so sure that the warming is caused by humans? That's another way of saying we don't know whether humans are causing global warming or not."*

But the review period was over, and the Third IPCC Assessment Report was ready for distribution within several months. Further revisions could not be considered at this late date.

To further confound matters, in early November 2000 an even

more pressing reason to provide the answer to the Jimmy Carter question about humankind's greenhouse gases being responsible for global warming had emerged. The incumbent Clinton/Gore administration had been very friendly toward this entire global warming investigation with its growing endorsement of the expanding anthropogenic "movement" during these years of uncertainty. But with the presidential election just over, a new era was about to begin under the leadership of George W. Bush and Richard Cheney, both of whom had strong ties to the fossil fuel industry. Thus, the inner core of climate scientists, while assessing the prospects of their human-caused global warming premise, became apprehensive about the future outlook and policies that might become instituted under this new leadership, particularly if it had not yet clearly been scientifically established that humans were indeed responsible for global warming.

So the best thing the inner core could do would be to "clarify" the IPCC Third Assessment Report and get the idea across to the public that there was no doubt that the human use of fossil fuels was responsible for global warming and climate change (regardless of what the report actually said) before new policies could be initiated by the about-to-start Bush/Cheney administration. After all – who reads long boring reports?

The Message Gets Across

In January of 2001, there was a World Environmental Conference coming up in Shanghai, China. (Coincidentally it occurred at the very time of the U.S. presidential inauguration). The highly respected British climatologist Sir John Houghton was scheduled to give an introduction to the new IPCC Third Assessment Report that was about to be released. Sir John had been a strong believer in the anthropogenic position for some time. He had "made magic" with the flat-handled hockey stick graph ever since he first saw it. Now, perhaps what he said and the way he said it, would be interpreted by the anthropogenically inclined media and the public to mean that humans, indeed, were responsible for global warming and climate change.

Reporting on the Shanghai conference in January 2001, a BBC news headline proclaimed:[72]

Human Effect on Climate Beyond Doubt

Thus, despite the uncertainty, a pivotal point in the global warming discussion was reached. Humankind was now reported to be responsible "beyond doubt" for global warming.

Here, one begins to see what might be termed the "slight departure" of the statements by the increasingly empathic journalistic community from what the inner core of climate scientists apparently actually had said. The BBC reporter at the Shanghai Conference, environmental correspondent Alex Kirby, stated: "*One of the world's leading climatologists says he is certain that human activities are warming the world.*" Kirby stated that Sir John said: "*I don't feel personally that there can be any doubt about the human effect on climate. The evidence is certainly sufficiently strong for countries to take action based on what we've said.... It's a highly authoritative report... the meeting ended with the report being agreed unanimously... it's probably one of the most peer-reviewed documents you could ever find.*" This kind of masterful verbal confabulation led to headlines such as the one above ("Human Effect on Climate Beyond Doubt"). There was no effort to correct this slightly impaired statement – after all, "66 percent sure" is hardly "beyond doubt". And this kind of scientist/journalist synergy became commonplace from this point forward in reporting on the field of climate change. Legions of journalists succumbed to the "informational cascade" that became the baseline for the anthropogenic global warming narrative. And the journalists enrapturement grew as the "movement" grew.

Also playing a role in all this was the previously stated conviction by members of the inner core that it was "for the common good" if everyone thought it was an established fact that humankind was responsible for global warming so that the public would take action to curb carbon dioxide emissions – for even though there was no confirming evidence yet that this notion might be correct – it later might turn out to be true[73]. And if it later did turn out to be a valid notion, the consequences might be disastrous if the public did not take action now.

Such scientific/journalistic synergistic actions allowed scientists to maintain a degree of scientific integrity while also allowing the journalists to be unchallenged as they led the public to a conclusion that was a departure both from the actual truth and from what actually had been stated by the scientist.

The casually uttered "without doubt" statement by Sir John was then repeated by the press throughout the world and became a rallying point by enraptured environmental activists. Many not-closely-affiliated with the anthropogenic movement but highly respected scientists from other fields of science also "picked-up" on these words and, as good followers of the *experto credite* wisdom also parroted these sentiments. The anthropogenic global warming movement was gaining momentum. The science behind the anthropogenic premise was still only somewhere around the *supposition* level of certainty. Yet the IPCC report had elevated the certainty of the anthropogenic position well beyond where it belonged. And now the media was elevating it even more – into the *we are certain* category.

This, once again, illustrated that we are human beings and we do have our share of unconfirmed convictions as well as communication misunderstandings. Apparently, everyone involved thought they were doing the right thing. The public was being told that a concept was true and established – while it was really unconfirmed or perhaps even completely incorrect.

Thus, by this point (early twenty-first century), much of the world thought that the scientific community had "proclaimed" that humankind was responsible for the global warming and climate change we were witnessing. While really, the scientific community had not actually made any such proclamation. The media apparently had reported what they perceived scientists had said – and the scientific community simply had let it stand without correction. *Thus, the public began to believe the scientific community had concluded "without doubt" that humankind was responsible for global warming and climate change. Sometimes this point in the evolution of the anthropogenic climate change illusion (January, 2001) is called the "proclamation of anthropogenic cause".*

Just to illustrate this important point of inflection, Al Gore's histrionic book *An Inconvenient Truth* was published in 2007. It, along with its derivative motion picture with the same name, was very helpful in convincing millions of people of the severity of the "Climate Emergency" we humans presumably had created for ourselves. On the first page of the introduction, Al Gore states:

> *"I have learned that, beyond death and taxes, there is at least one indisputable fact: Not only does human-caused global warming exist, but it is also growing more and more dangerous, and at a pace that has now made it a planetary emergency".*

With this notion now an established fact (at least to Al) the book spent the next 300 pages telling the reader all the horrible things that would happen to us if we continued to use fossil fuels. And it was all based on an unconfirmed supposition that apparently he and millions of others thought was based on rock-solid science.

It is unfortunate that this conclusion of human culpability was drawn at that time, for as it later turned out the conclusion would have been quite different if it had been delayed until the science – particularly of the sun's slowly (several hundred year) rhythmically changing magnetic field and its interaction with the earth's atmosphere and clouds – could have been more deeply explored.

This *proclamation of anthropogenic cause* of January 2001 was indeed a pivotal point in the making of the widely-believed intellectual illusion that human beings and their fossil fuels and greenhouse gases are responsible for global warming and climate change when, as will become more evident further on in this book, neither human beings nor fossil fuels and their resulting greenhouse gases are the principal causes of global warming or climate change.

Chapter 4
The Anthropogenic Global Warming "Cause"

Fortune favors the bold.

— Ennius, 200 BC

TIME MAGAZINE'S THEN-MANAGING-EDITOR, RICHARD Stengel was the first popular-magazine mogul to hop on the anthropogenic bandwagon. The cronyism that was developing between journalists and the advocates of the anthropogenic global-warming cause began to show when Stengel said in an interview on MSNBC: "*You [journalists] have to have a point of view... You can't always just say on the one hand, on the other, and you decide.*"[74]. The public was apparently as ready for a definitive answer to what was causing global warming as the scientific community was. This investigation and discussion had been going on for two decades with huge research expenditures – yet the scientists had not yet been able to convince much of the public of the conclusion that the environmental community thought they "knew to be true". So the altruistic journalists decided to help.

Newsweek, National Geographic, Popular Science, The New Yorker and in fact all the popular consumer magazines demonstrated their empathy – and their sympathetic inclination – by following suit.

The social atmosphere was ripe for acceptance of this massive failure by humankind. The environmental movement had been pointing out humankind's many infringements on our natural world. And now virtually everyone seemed to want to believe that human beings were responsible for global warming and climate change – based, quite simply, on the obvious facts that the world seemed to be getting warmer, and that humans were pouring huge quantities of a greenhouse gas (carbon dioxide) into the atmosphere. Small pockets of scientific dissent were quickly dismissed by the journalistic community as minor impertinences by those awful "deniers". Virtually everyone who did not have a stake in the fossil fuel industry seemed to be either blaming humankind or at least acknowledging "*it could be true that humans are to blame*". Very few in either the scientific community or the general public seemed to comprehend that the amount of carbon dioxide in the atmosphere was so tiny that even a hugely significant increase would have only a tiny impact on global temperature.

Change of Phraseology of Global Warming.

The transition from the phrase "global warming" to "climate change", and more recently to "climate crisis" and finally to "climate emergency" was advanced by global-warming enraptured environmentalists who became increasingly concerned that in each instance, the current phrase was not alarming enough to motivate the public to expeditiously implement the changes and accompanying sacrifice that might be required to stem the flow of greenhouse gases into the atmosphere. Essentially this shows the frustration felt by that community at the lack of progress toward a goal they felt was imperative to rapidly achieve. The change of phrase was another one of the things they could do to try to increase the sense of urgency in what was an extremely slow-moving process – the hoped-for conversion from fossil fuels to "renewables".

James Hansen, head of NASA's Goddard Institute for Space Studies, was without doubt the most evangelical advocate of this widely-held anthropogenic conviction in the U.S. scientific community. He gave hundreds of congressional, press, and TV interviews over the next several years blaming the human use of fossil fuels for global warming. For a guy who was director of an important research facility, that's a lot of time to spend on press interviews – illustrating how important he felt the message was. Jim Hansen, with his public-awareness efforts, had become the prime climate science expert advocate for the anthropogenic cause.

On the political and celebrity front, Al Gore, originally alerted by Roger Revelle's thought-provoking but cautious approach to the subject at Harvard, had quietly but effectively worked as a junior congressman from Tennessee to present anthropogenic advocates such as James Hansen to testify before various key congressional committees. Gore had by now become convinced that human beings were causing global warming. As time went on, he became a powerful political/celebrity/environmentalist advocate for the anthropogenic cause, and developed his popular lecture, histrionic book[75], and dramatic documentary movie[76], all of which aimed at alerting the general public to the many dangers of human-caused global warming and "climate change" as it was then frequently called.

Sir John Houghton, the aforementioned prime British scientific/celebrity anthropogenic advocate, regularly used the hockey stick graph as his favorite validator and gave many popular lectures. To an impressionable public, Sir John added considerable prestige to the notion of anthropogenic global warming.

In terms of influence, Hansen – the American scientist, Gore – the celebrity environmental advocate, and Houghton – the British scientific figurehead, were the "big three" in influencing the wider scientific community and the public that human beings were causing global warming. They were backed up by the rest of the inner core of anthropogenic climate scientists, many of whom seemed to be becoming enraptured by the notion of anthropogenic global warming. Nearly all became such strong believers in the anthropogenic doctrine that they were no longer able to be influenced by the emerging contrary scientific revelations as the subject was more deeply explored – an unfortunate place for scientists to be.

To top it all off, the Norwegian Nobel Committee decided in 2007 to bestow upon the IPCC and Al Gore, the Nobel Peace Prize for their work[77]. Just what global warming has to do with peace remains a mystery to many, but it seemed to be a form of validation for a group of distinguished Scandinavians to bestow on someone a prize – even a malapropos one – based on the unestablished but the now-becoming widely accepted notion that humans cause global warming.

This was all followed in 2009 by the U.S. Environmental Protection Agency (EPA) with its "Endangerment Finding"[78], which concluded that greenhouse gas emissions posed a threat to human health and must be regulated under the Clean Air Act (which will be discussed later, in chapter seven).

The Elevation of Certainty

The scientific paper that Kiehl and Trenberth wrote in 1997, [K&T '97], which the inner-core used as their defining basis and thus became the theoretical foundation upon which the whole concept of anthropogenic global warming was based, was published as an update on the energy balance of the earth. For this use, the breakdown or "partitioning" of the greenhouse gases did not really matter. Unfortunately, the quantified "partitioning" part was used (by others) to calculate climate sensitivity – traditionally defined as the number of degrees Celsius global temperature would rise as a result of a doubling of atmospheric CO_2 content. This led to the determination that humankind was responsible for global warming. This K&T '97 paper contained unverified assumptions, uncertainties, and issues, as well as acknowledgment of that by the authors who stated:

"The values put forward are reasonable but clearly not exact. The purpose of this paper is not so much to present definitive values but to discuss how they were obtained and give some sense of the uncertainties and issues in determining these numbers..."[79]

But the quantified results: the partitioning of 60 percent water vapor and 26 percent CO_2 temperature forcing in the clear-sky part of the atmosphere was able to be used to calculate a climate sensitivity figure of 1°C rise in global surface temperature for a doubling of atmospheric CO_2 (which became 3°C after application of the water vapor feedback amplification factor). This K&T '97 paper, despite its reservations of uncertainty, but with its quantified conclusion, was then cited by climate scientists as the theoretical basis for their conclusion that humankind was primarily responsible for the global warming we were witnessing. It was interpreted by other scientists in the peripheral scientific community to provide an affirmation of what they wanted to believe and had been saying all along. So quickly the certainty of these parameters began to be elevated with the words: "...easy to calculate and uncontested"[80]. Later, this was further elevated to: "... easy to calculate and undisputed"[81], while the U.N. changed the word to: "unequivocal".

With this kind of treatment (that is, assuming a low-level-of-certainty "supposition" to be "without doubt"), the peripheral scientific community was off and running despite the lack of any solid scientific evidence. They finally had what appeared to be a theoretical consequential figure for anthropogenic global warming. Something that they thought would support their peripheral research and could be attributed (however tenuously) to their "expert" colleagues. Thus, the narrative spread widely beyond just the climate scientists into the peripheral scientific community, and then into the environmental community, the policy-making community, and the public.

But scattered scientific skepticism remained. When later pressed to the wall by penetrating questions, Trenberth stated: "*This is just one scenario....*" Well, "*just one possible scenario*" combined with the already stated: "*clearly not exact*" and "*give some sense of the uncertainties and issues in determining these numbers*" is far different from "*uncontested*" or "*undisputed*" or "*unequivocal*" which are the words the peripheral scientific community, the environmental community, and the media were running with. Sometimes, we as human beings get ourselves into these situations for today we are told that 90 percent (or sometimes the "overwhelming majority") of the scientific community is convinced of something that the authors (Kiel and Trenberth) of the foundational scientific paper upon which this whole premise was based[82] thought was "clearly not exact..." and there were "uncertainties and issues" remaining, as well as that "it is just one scenario..."

Chapter 5
The Sleeping Giant

God is usually on the side of the big squadrons...

— Roger de Rabutin, 1677

THE FOSSIL FUEL INDUSTRY had been building over several hundred years starting with the use of coal for the steam engines of the Industrial Revolution in Britain in the late 1700s. This slowly increasing use of fossil fuel proliferated into Europe and America in the 1800s.

The major sources of man's energy up until that time had been muscle (man and his pack animals – primarily the horse, donkey, ox, llama, camel, and elephant); wood; wind; water wheels; and fish, animal, and vegetable oils for lighting.

Wood was becoming scarce in many parcels of dense population in Europe. For example, parts of chilly Scotland had been particularly deforested and the affected people had started becoming dependent on peat (which fortunately was conveniently situated on the ground surface) for cooking and heat.

Wind was another easy energy source both for the many sailing vessels of the world and also the windmills of Europe (and elsewhere – such as China). Windmills were often used for pumping water (as in Holland which was reclaiming land from the sea), or for grinding grain, where the grain could be stored until the wind blew.

Water wheels were used throughout the world mainly for grinding grain, but later-on (and more often than commonly recognized) for the hammering of metal such as iron by smiths, as well as for other repetitive-operation uses.

The Brief Whale Oil Era

Various fish, animal, and vegetable oils were already being used primarily for lighting, when the superiority of whale oil was recognized. A favorite feeding ground for humpback whales was Stellwagen Bank between Cape Cod and Cape Ann in New England. As the carcasses of dead whales washed ashore, local Wampanoag Indians had been extracting oil from the blubber. Ingenious Yankees, with their superior shipbuilding capability, began to extend the practice to harvesting live whales and extracting the oil right on the ship. Reports of the superiority of whale oil as fuel for lamps quickly spread. This led to the whaling industry that, within a matter of a few decades, grew to over 600 sailing vessels in the Yankee whaling fleet by the mid-1800's. Whales became increasingly scarce as the demand increased, and the price of whale oil escalated from thirty-five cents to eight dollars per gallon in just a few years as the whalers had to travel to the far reaches of the world for their quarry.

The rising price of whale oil combined with the much lower cost and increasing availability of kerosene (derived from petroleum) ended the brief whale-oil-for-lighting era.

Kerosene for Lighting

Petroleum (a fossil fuel), had enjoyed small-scale use from seeps and dug wells for many years. In the 1200s, Marco Polo commented on a *"fountain of oil, used not as a food, but as an unguent and for burning in lamps and heating (and distributed by caravans of camels)"*, as he passed through the Caucasus *"north of Armenia"* on his classic journey[83].

More recently, petroleum entered the full-scale picture as the Nobel brothers in 1844 in what is now Azerbaijan (very near where Marco Polo's fountain was), and Drake in western Pennsylvania in 1859 began to drill wells in places that had vast reserves of the precious substance. Soon kerosene for lamps was at thirty-five cents a gallon and the number of ships in the Yankee whaling fleet quickly dropped by more than 90 percent (fortunately for the poor whales who were being decimated).

Waterpower

Waterpower from flowing rivers provided energy for the growing textile industry in New England. Much of the waterpower of the state of Rhode Island, where the U.S. Industrial Revolution began, was harnessed from small rivers such as the Blackstone, Wood, and Pawcatuck which were easy to dam but didn't have much power at each dam site. This led to the "mill village" system wherein small factories were sited at each source of energy – a series of dams over the entire length of the river. The water from each dam spilled into the upper end of the mill pond of the next dam – thus using virtually all the energy of the river[84].

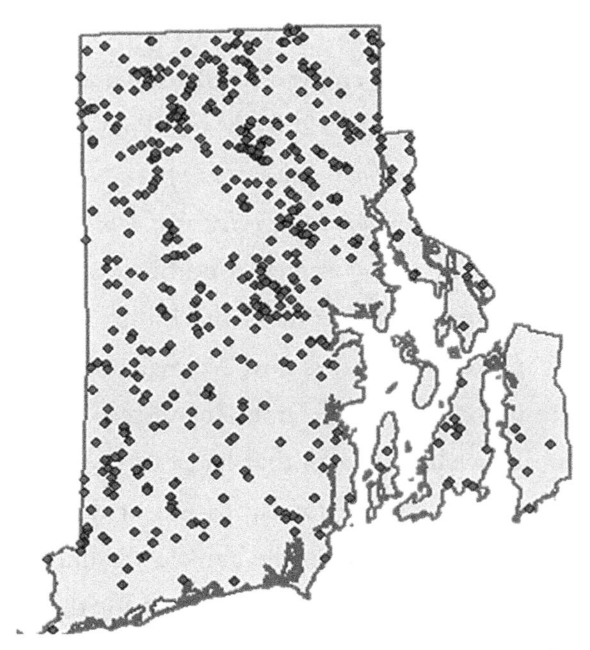

Graphic 5.1 Historical map: **Dams in the State of Rhode Island.** R.I., D.E.M.[85].

Initially, in the New England States, many of the industrial sites with waterpower near the sea coast were already producing metal parts for the wooden shipbuilding industry located there. Thus bars, bolts, nails, screws and other fastening parts were being produced. That metalworking capability was welcomed and effectively used by the fledgling and fast-growing textile plant equipment suppliers for the required machinery. Many of the early textile mills were thread mills since thread-spinning was the most labor-intensive step in the textile process.

But thread making is far more than just spinning. Perhaps the most difficult part during the start-up phase was developing and synchronizing the raw material preparatory steps which required a series of machines in order to prepare the raw cotton or wool for the spinning operation. Thus carding, and then drawing, and roving frames had to be perfected as well as the more dramatic spinning frames.

Samuel Slater, a knowledgeable emigrant from Derbyshire, England was called the "Father of American Manufactures"[86] for his skill not only in bringing the proper technology for the individual machines to the United States from England – but providing the overall co-ordination of the entire thread-making process. In addition, giant contributors such as the financial and organizational power of the merchant and banker Moses Brown and the machine-making and tooling capability of George Wilkinson should not be forgotten. Together with Slater, they produced the first truly successful textile mill in America at Pawtucket, Rhode Island in 1792. *We must recognize and give credit to these men and their associates for starting what was to become the great American Industrial Revolution, a transformation of society unlike any other.*

As time went on, essentially the power of the entire system of rivers and tributaries in the state of Rhode Island was tapped for industrial purposes (mostly textile). See graphic 5.1. Each mill village was populated by families at the factory site who lived in either private or company homes. Everybody worked at the mill. Many of the bigger villages had a school and store – all provided by the company. School, by the way, was mostly on Sunday since the children worked in the factory six days a week, twelve hours per day. (Some progressive employers cut Saturday's hours to six). The most prized workers were boys from eight to twelve years old (called "doffers") who could quickly remove and replace bobbins, spools, and spindles, as well as squeeze under or around almost any machine to tie a broken thread or release a snag. In one instance the superintendent of the mill was a twelve-year-old boy, having worked his way up from the age of eight (and, incidentally, also being the son of the owner)[87].

As demand increased, the need to expand the output of the

mills was held back by the lack of sufficient water flow. So the switch from the water wheel to the water turbine allowed increased capacity. Later, of course, the steam engine came along.

We must acknowledge the work of other early industrialists such as Francis Cabot Lowell who started the Boston Manufacturing Company in Massachusetts, the first completely integrated textile mill (from thread to finished cloth) in the United States some two decades after Slater's first thread mills in Rhode Island. These larger integrated mills usually used young ladies aged 18 to early 30s for labor (mostly just off the farm), who lived in chaperoned boarding houses. These girls would normally return home to get married after a few years of "living a different life". Sometimes they would wait too long, returning to their rural homes from the spinning mills after all prospects of marriage had passed them by, and would be termed "spinsters".

Back in England more and more deposits of coal were discovered in tandem with the improvements in the steam engines. And once some of the technical problems such as water leaking into the mines were solved there never seemed to be a shortage of this available commodity to fuel the burgeoning energy requirements of the fledgling Industrial Revolution. Fortunately, Thomas Newcomen had developed the first crude atmospheric steam engine back in the early eighteenth century and pumps powered by these innovative devices were able to drain the water out of the mines faster than it leaked in, and keep the mines open.

Hargreaves, Arkwright, and Crompton invented and improved the mechanical spinning process between 1764 and 1779. The spinning machinery was at first hand-powered and then horse-powered and then water-powered. These early spinning machines evolved from the spinning jenny, to the water frame, to the spinning mule – a major segment of which spectacularly rolled back and forth across the floor during the spinning process.

As the demand for thread increased and the powered carding, drawing, and roving frames and spinning jennies and frames proliferated, it became necessary to automate the weaving process in order to keep up with the growing quantity of available thread. Kay's hand operated "flying shuttle" was a significant first step, but it still required an operator to throw the shuttle.

Then, as powered weaving looms became available, there was a growing need for more sources of power particularly in Britain where water power was less available than in the United States. James Watt's constantly improving more advanced double-acting steam engine with external condenser filled that need, and the steam-powered Industrial Revolution was off and running. Coal-powered steam engines became the new workhorse for this fledgling textile industry. After some years of technology advances on both sides of the Atlantic, coal-fired steam engines culminated with the gigantic "factory size" engines of George Corliss from Providence, Rhode Island.

Graphic 5.2. Historical etching: **The Corliss Four-Valve Steam Engine at the 1876 Philadelphia Exhibition.** The Engineer's Handy Book, 1884[88]. This 45-foot tall, 1400-hp, 650-ton behemoth powered the entire 13-acre exhibition display of 8,000 machine tools through five miles of shafting and one mile of overhead line belt. The 56-ton flywheel revolved at 36 rpm without vibration. The single operator spent most of his time reading the newspaper, while spectators gawked.

The 1876 Philadelphia Exhibition

This early "World's Fair" demonstrated the potential of America and its industrial development with not only this mammoth steam engine, but also: a Bratton hydrocarbon engine (a precursor to the coming automotive age); an ammonia compressor for refrigeration and ice making; an Otis steam-powered elevator; a Baldwin steam locomotive; an R.M. Hoe rotary cylinder printing press (that allowed the use of continuous rolls of paper); a Seth Thomas grandfather clock; a Wallace-Farmer electromagnetic generator (a harbinger of the coming age of electricity); Alexander Graham Bell's telephonic communication device; Waltham Watch Company's automatic precision screw machine (forecasting the era of close tolerance manufacturing and standardized interchangeable parts); a Remington typewriter; Heinz ketchup; Hires root beer; and many others including, in the international hall, a Krupp blunderbuss cannon – the largest in the world. Each of these items and the technology involved in its development was, in itself, revolutionary. Collectively they marked the beginning of a transformation in the very essence of civilization[89].

Back in Great Britain, James Watt's more modest-sized steam engines had become self-tending enough to be used for transportation power – and thus the steam train (and later the steamboat) became possible.

Labor supply expanded with the growing population, creating a surplus of potential workers. It was easy to recruit workers to endure the hostile environment of the coal mines as well as to operate the carding and spinning frames and weaving looms for 72 hours per week. The growing population fed the growth of industry and the growth of industry fed the growing population.

Coal, instead of wood, began to heat more and more houses in winter as well as to power the steam engines that ran the factories year-round. By the mid-nineteenth century, the transition from wood to coal as the primary energy source in the western world was well on its way.

Fulling, Walking, Tucking.

One of the important steps in producing woolen goods was "fulling", which is a cleansing step. It was also called "walking" and "tucking". Who doesn't know someone named Fuller, Walker, or Tucker, or even Spinner, or Weaver?

Graphic 5.3. Historical photograph: **The DeWitt Clinton.** Henry Ford Museum, Detroit, Michigan[90]. First steam train publicly used in New York State, 1831. Note the wood fuel in the barrels. Although coal was available, wood was apparently still cheaper than coal.

In North America, as electric power developed in the late nineteenth and early twentieth centuries, more and more of the coal transported by steam trains was used to generate electricity, and the electricity was transported to the centers-of-use by high-voltage transmission lines. This meant a more unnoticed, yet rapidly increasing use of coal for electrical power. As the twentieth century matured, the publicity was focused on the awe-inspiring new large hydroelectric plants – Grand Coulee, Boulder (Hoover) Dam, and the TVA dams, but the backbone of the "transformation to electricity" was in the many (ultimately over 500) coal-fueled power plants that were located all across the country.

Graphic 5.4. Historical advertisement. **Winton Motor Carriage**. First auto advertisement[91].

Petroleum, at first utilized mainly as kerosene for lighting, began to have an even more dramatic increase in use than coal, as the new internal combustion engine entered the transportation market. Automobiles, trucks, and airplanes began to proliferate and ships and trains converted from coal to petroleum. Petroleum liquids powered not only the newly developed internal combustion engine, but also the Diesel engine and then the turbojet as they came on-line. Thus, petroleum liquids quickly became the favored fuel for any form of transportation (as kerosene for lighting gave way to electricity).

Natural gas, although much more difficult to store than petroleum liquids, found more and more uses as distribution and storage facilities expanded, and they have continued to expand until today. Currently our fossil fuel use in the United States is close to being evenly divided between coal, petroleum, and natural gas – although natural gas is rapidly displacing coal in many areas – such as electricity generating power plants.

In the United States, most of this fossil fuel energy increase in use was relatively steady, uneventful, seamless, and not that obvious. It became expected that there would always be a readily available supply of reasonably priced fuel of all types – natural gas, coal, petroleum, and electricity. There was little fanfare upon the opening of a new coal-fired electric power plant and they were being opened at an ever-increasing rate throughout the country as the twentieth century wore on. High-voltage transmission lines went from the remote sources of generation (dams, large coal-fired generating complexes, and nuclear-powered stations) to the centers of population and use. Mid-voltage transmission lines reached out from the centers of population to the suburban areas and to the rural population who welcomed the arrival of central-power electricity. Now refrigeration replaced the "ice box", electric lights replaced kerosene lamps, and the water well could be pumped by electric motor instead of by a windmill or by hand. *The electric generating capacity of the United States doubled every decade of the twentieth century up through the 1970s and most of it was fueled by coal*[92].

By the turn of the twenty-first century, the Powder River basin coal formation in eastern Wyoming and southeastern Montana, with four enormous surface mines was shipping out about 80 trains per day, each with about 100 hopper cars, each of which was filled with 100 to 120 tons of coal, to middle-American power plants[93]. Every day! That's a lot of coal.

There in eastern Wyoming, as far as the eye could see, were the windswept fields of native switchgrass that supported sparse transient flocks of pronghorn antelope. But underneath those fields was pure energy. About six feet of topsoil was scraped off and piled to the side. Then 50 feet or so of subsoil was removed with huge excavators and piled to the other side.

Then a wide expanse of an 80-feet deep (or so) vein of pure clean coal was extracted by huge power shovels and conveyed a short distance by heavy-duty haul-trucks to the cleaning and processing plant and then on to railroad hopper cars for travel to the power plants in some 30 states on any of four of the country's largest railroads. As the excavating process moved forward the backfill took place, with the subsoil replaced on the now-exposed next layer of subsoil, and the topsoil replaced on that. With appropriate switchgrass reseeding, the pronghorns would soon return and the windswept fields looked much the same as in the virgin state (except that the level was 80 feet lower). This environmentally friendly treatment of the terrain demonstrated that the coal mining industry could indeed be a good neighbor (having perhaps learned their lesson earlier and elsewhere). It's a little more costly to finish things off this nice clean way but it's well worth it from an environmental (and public relations) viewpoint.

After complaints about "smoke" and "acid-rain" in the Northeast, large coal-fired electric generating plants, (that used mainly Appalachian coal which has a higher sulfur content than Wyoming coal), became equipped with exhaust scrubbers. Sometimes they used a combination of the very clean yet economical electrostatic precipitators to remove the larger particulates, and the water and powdered "lime" scrubbers for the finer pollutants and sulfur oxides. These plants did not belch black smoke, and thus were less obtrusive than those that did not treat their exhaust. They did give off white steam, but an observer could see it readily dissipate harmlessly into the air. And they did give off carbon dioxide – lots and lots of carbon dioxide, which is both colorless and odorless. The "lime" from the scrubbers is sometimes combined with the sulfur in the coal exhaust to produce synthetic gypsum. This is then used for a variety of purposes. Four National Gypsum Company plants produce wallboard for the construction industry exclusively from synthetic gypsum produced from the exhaust of coal-burning power plants[94]. And, of course the acid is largely removed and the air neutralized, by the removal of the sulfur.

Even today many people confuse the black particulates of smoke with the colorless pure gas – carbon dioxide. Some of our Asian fellow-world-community members such as China and India could learn from our playbook because these kinds of air scrubbers, while expensive, transform a former heavy polluter into a much better neighbor – and even give a dividend in the form of very useful wallboard. There is a price to pay for clean air but it is worth it. In Asia, the air pollution problem is more complicated than ours because they have multiple other sources of smoke – such as their coal burning heavy industry; the domestic and commercial use of coal, dung, and wood for light industrial and commercial processes as well as home cooking and heating; their custom of field burning grain crop straw; and their frequent fireworks displays, as well as unfiltered and uncatalyzed internal combustion engines. Today Beijing, from an air pollution perspective, might be equated to Pittsburg in 1944.

These United States coal-fired electric generating plants often went relatively unnoticed by the general public – including the scientific community, who had no more reason to think about how dependent we had become on fossil fuels than did any other person on the street.

Likewise, the increase in petroleum use was equally steady and undramatic. Today there are more than 900,000 operating oil wells in the United States which partially supply some 265 million automobiles (with some help from imported oil and ethylene from corn).

Between the exhaust "scrubbers" in coal-burning power plants and the catalytic converters in automobiles, in the U.S. we've come a long way in cleaning our air. However, we still have a long way to go in some areas – such as the mountain-encircled Los Angeles basin and other densely populated cities with uncooperative atmospheric conditions. The continued increase in electric vehicles in these "air-pollution problem" areas could be a big help if the fossil fuel electric generation (and its attendant exhaust) required to recharge the batteries in those vehicles is located outside the problem area – where there are more favorable atmospheric conditions – and thus where the exhaust can be more readily controlled and dissipated. Today, much of the electric power used in the Los Angeles basin of California is coal-generated in a huge complex in the desert area of Utah where the exhaust harmlessly dissipates. Thus, the air pollution problem in Los Angeles is largely from transportation and industry. Near total conversion to electric vehicles by residents and government entities would go a long way toward solving this particular problem (although it would require the expansion of the electric generating complex in Utah or in some other atmospherically suitable area).

At the turn of the twenty-first century some nine percent of United States electricity was hydroelectric power, and 19 percent was nuclear, but over 70 percent was provided by fossil fuels (split mostly between coal and the fast-gaining natural gas)[95].

Chapter 6

The "Cause" Becomes a "Movement"

Let's follow him, he seems to know where to go.

— Ancient animal instinct.

HUMAN BEINGS ARE GENETICALLY predisposed to becoming enraptured by a new and exciting concept, cause, or movement – it seems to fulfill a basic human need. And if it is a righteous cause it is even more alluring. As Princeton physicist William Happer states about the climate change enrapturement: "What better cause than saving the planet?"[96]

The early 2001 reporting in the media of the proclamation at the Shanghai conference that humankind was responsible for global warming "beyond doubt" was a major point of inflection in the march to answer the question: *Is human use of fossil fuels causing the earth to warm?* For the question illuminated something more powerful than science: The premise of anthropogenic global warming was becoming a full-fledged "righteous cause" or "movement". Captivated scientists-turned-activist were leading the charge – and many environmental activists as well as journalists, politicians, and members of the general public were now firmly convinced of the nobility of their cause: "*We must stop using fossil fuels or we are all doomed*".

Unfortunately, to some the anthropogenic global warming "cause" was becoming more important than the science that backed-up that cause – and when that happens, sometimes we reshuffle our moral priorities emphasizing the "cause" and de-emphasizing the scientific integrity that might interfere with the affirmation of our newfound ideological imperative. But, after all, we are only human.

The inner core of anthropogenic climate scientists had committed themselves to the anthropogenic supposition early-on, well before there was really adequate evidence for them to make such a commitment. Accordingly, their reputations were on the line. So for many of them, promulgating the cause had now became more important than whether humankind was actually responsible for global warming. Who wanted to look into the unanswered question: "Are humans causing the earth to warm by using fossil fuels?" when "everyone" already knew that they were. So the best course for the now-enraptured coalition of the anthropogenic global warming environmental, scientific, and media communities was to simply keep stating that humankind was responsible for our changing climate while ignoring any and all contradictory evidence – and calling anyone who even hinted at such blasphemy a "denier".

The IPCC Third Assessment Report was issued in early 2001 as it had been written[97] (the IPCC was about 66 percent sure that humankind was responsible for global warming). A few peripheral scientists and impartial scientific journalists perused the report looking for the research "finding" that would justify the sudden conclusion so widely reported in the media that climate scientists were now sure "beyond doubt" that humankind was responsible for most of recent global warming that journalists reported had been "declared" in Shanghai. No such finding could be found in the third report.

The 66 percent sure *judgmental assessment* was a much higher degree of certainty than the evidence in the report warranted. But the anthropogenic inner core wanted to see an even higher level of certainty. After all, the press had stated that the scientific community was sure "beyond doubt" that humankind was responsible for global warming while the IPCC report was only 66 percent sure. As stated earlier, it was at this point that what was *being reported* about the certainty of the premise of anthropogenic global warming began to diverge from factual reality and started to become much more certain than the scientific evidence really demonstrated or the originating scientists had concluded. This led to a misunderstanding on the part of peripheral scientists as well as the journalistic, political, environmental, and general public communities who now all had a far more certain belief in human-caused warming than the evidence warranted.

But with the clear disparity between the IPCC report and what the press was stating the inner core of climate scientists believed, it became necessary to put the two allies (the IPCC summarizers and the anthropogenic inner core of climate scientists) back on the same page. We couldn't have the "experts" saying one thing and the "summarizers" another. So the experts, primarily by remaining silent, held to the story that there was no doubt that humankind was the prime cause of global warming and simply let the IPCC summarizers catch up by upgrading their degree of certainty in the next assessment report. It should be noted that there was no absolute or pure scientific basis for the proclaimed conclusion that humankind was responsible for most of global warming. The basis was only a *judgmental assessment* – much of which was derived from an *inferred conclusion* that was to a large extent based on a *preconceived supposition* that was backed-up only by incomplete and, at best, low level-of-certainty (or suggestive) theoretical, observational, and computer-model evidence, and psychologically augmented by the growing number of enraptured members of the movement.

So, as we moved into the twenty-first century, the IPCC public-awareness section – assisted by a very friendly journalistic community, many of whom had become suitably enraptured themselves after a dozen years of persistent propaganda coming from the dedicated IPCC *public awareness* or *public opinion guidance* section, helped to spread the anthropogenic message. They frequently repeated "The case is closed", "Climate Change is a fact"[98], "The time for debate is over"[99], and "The science is settled"[100]. Skeptics and contrarians were regularly termed "deniers" not only by the anthropogenic environmentalists but also by the mainstream journalistic community.

And the broader peripheral scientific community seemed to be on board with their new-found: "*The global temperature rise for a doubling of atmospheric CO_2 will be 3° C*" climate sensitivity calculating factor, which quickly became widely used for downstream research as if it was an accepted and confirmed law. In addition, the IPCC Detection and Attribution summarizing section had now acceded to the inner core of "experts" and had accepted their *global warming is human caused* "*without doubt*" statement without any further supporting scientific evidence. And they ignored virtually all evidence to the contrary that was emerging in unorganized and uncontextualized bits and pieces.

At this point (the beginning of the twenty-first century) there was no "proof" of anything in this field – and the "experts" must have recognized the uncertainty. They could only hope that the evidence would become more convincing as time went on since there was not a definitive case demonstrating that humankind was responsible for the global warming we were witnessing. It was the previous commitment (made in the IPCC First Assessment Report of 1990) that the inner core had made to the National Science Foundation that they would have an answer within a decade – as well as a need to preempt any new philosophies that might emerge as a result of the recent presidential election – that caused the inner core to make a pronouncement as definitive and as early as they did. Essentially, they were forced to make a decision well before there was any real evidence of certainty. And now if things would just become more and more certain as additional evidence unfolded, the inner core and the IPCC summarizers, finally on the same page, would be all set. Or apparently at least they so hoped.

Obviously, the feeling among the scientists at the center of the inner core at this time was still that "Even if we don't have a solid scientific basis for demonstrating that humans cause global warming, the fact that it *could be true* is the important part. So, if we must allow the notion that *it is true*, to be instilled in the minds of the general public in order to get them to take steps to curtail further emissions of CO_2 into the atmosphere (just in case that might later turn out to be true) then that's the way it will have to be"[101].

This was really a crucial time in the anthropogenic global warming controversy. It was obvious that the leaders of the "movement" (the inner core of climate scientists as well as an increasing quantity of celebrity advocates) had no real grasp of just how dependent the world had become on fossil fuels. They believed, as Sir John had stated at the Shanghai conference: "*The technology is there to do something about it. Energy efficiency, alternative energy sources, even carbon sequestration – none of them would cost very much... Drastic changes in lifestyle are not what we are talking about. There'll still be cars – but they'll be much more efficient...*[102] "

Idealists see where they think society should be going, without worrying about how they will get there, while realists worry very much about how they will get there while wondering why they even want to go there. In this instance the problems in attempting to wean the global community from using fossil fuels as the main energy source of a modern civilization that is thoroughly dependent on low-cost energy, were apparently completely unrecognized by the proponents of – what is turning out to be – an unattainable transformation of society.

It had taken over a hundred years of largely unplanned and underappreciated innovation, luck, pluck, trial-and-error, financial investment and risk – with attendant triumph and reward, or failure and bankruptcy – for the "correct path to success" to emerge. And this path was highly dependent on the massive use of fossil fuels. The ramifications of, and effort required, to replace this "super-massive energy movement" with something else was apparently beyond the comprehension of not just these "proponents of change", but also nearly everyone else too. The vastness of not only the direct fossil fuel industry itself but of its many dependencies and peripheral endeavors is simply too staggering for most people to comprehend. It is something that grew as we as a people increased in number, developed technologically, and as our civilization evolved, advanced, and matured.

An example of the trial-and-error nature of this technologically advancing aspect of human development that took place in the early twentieth century is the transportation explosion: Train? Ship? Bus? Airplane? Trolley? Auto? Within the auto category there were three main contenders: the internal combustion engine (Ford Model T); the steam engine (Stanley Steamer); and the electric motor (Baker,

Babcock, or Milburn electric cars). The primary conventional competition was the horse and carriage. These four "fought it out" for about three decades (with the horse and buggy slowly fading out and the internal combustion engine beginning to slowly take the lead in the competition. Finally, the gasoline powered automobile overwhelmingly won – no subsidies, no favoritism, just raw pluck, luck, and technical and logistic superiority – and the overriding kicker – a cost that consumers were willing to pay.

A side benefit evolved from this amazing contest. Henry Ford recognized that to sell as many vehicles as he needed in order to keep costs low, the workers making them had to be able to afford to buy one for their own use. So Henry Ford voluntarily raised wages to the point where workers could afford to buy a low-cost model-T. Not only did the model-T capture the automobile, market but the concept that: *workers, too, are "must" consumers* began to take hold. That forced other manufacturers to do the same thing. And the great American Industrial Revolution with the winning combination of: *low-cost production-line products and highly-paid workers who were now able to afford those products,* had shifted into high gear. This "New America" soon became the wonder and envy of the world.

The Power of the Mind

As we moved into the early twenty-first century, much of the mainstream journalistic community had come to rely on the IPCC's public awareness section to be the communication channel between the scientific literature and themselves. And with the limitations on discussion placed on the IPCC by its own charter and guidelines, the information that got through this highly-biased information filter and ultimately became shared with the media was heavily infused with the anthropogenic position. Thus, willingly or unwillingly, the media was to play a key role in the propagation of the anthropogenic cause.

Much of the reason that this anthropogenic global warming thinking was able to be spread so far beyond the inner core when it was so uncertain is perhaps explained by what social scientists call an "informational cascade". John Tierney explained this psychological process[103] by quoting the German philosopher Arthur Schopenhauer, who originally described how a notion or premise starts with one person and begins to "bandwagon" in chapter 3 of his perceptive mid-1800s collection of essays titled "*The Art of Controversy*":

> *"We should find that it is two or three persons who, in the first instance accepted it, or advanced and maintained it; and of whom people were so good as to believe that they had thoroughly tested it. Then a few other persons, persuaded beforehand that the first were men of the requisite capacity, also accepted the opinion. These again were trusted by many others… The remainder were then compelled to grant what was universally granted, so as not to pass for unruly persons who resisted opinions which everyone accepted, or pert fellows who thought themselves cleverer than everyone else.*
>
> *"When opinion reaches this stage, adhesion becomes a duty; and henceforward the few who are capable of forming a judgement hold their peace. Those who venture to speak are such as are entirely incapable of forming any opinions or any judgement of their own, being merely the echo of others' opinions; and nevertheless, they defend them with all the greater zeal and intolerance … there are very few who can think, but every man wants to have an opinion; and what remains but to take it ready-made from others…*

"Since this is what happens, where is the value of the opinion even of a hundred million? The opinion in the end being traceable to a single individual."[104]

Schopenhauer, more than a century ahead of the fact, foretold with astonishing vision, the psychological process of the cascading acceptance of the still uncertain notion of anthropogenic global warming and its accompanying movement.

A more current example of the "group think" or "herd effect" phenomenon is described in Irving L Janus' *Victims of Groupthink*[105]. S.E. Sondergard points out that the contributing members of the IPCC exhibit classic groupthink in every respect[106] (as do the "believing" members of the anthropogenic global warming "movement").

As it now applies to climate change, the popular wisdom has become so accepted within the scientific community that it is a career risk to question it (bringing to mind the "adhesion becomes a duty" of Schopenhauer, or what we now popularly call "*go along to get along*" or "*go with the flow*"). Skeptical scientists become ostracized (and cut-off from career-necessary financial grants), and public debate and the research agenda have become dominated by the overwhelming conventional wisdom. There simply is no room for disparate views. Thus, it is no longer necessary for members of the "movement" to counter, or even address adverse findings. The weight of inner core scientific and environmental activist opinion is so strongly in favor of the foregone conclusion that contradictory data is usually dismissed as "denialism" or simply ignored with impunity[107].

Finally, we have gotten to the point where key people are embracing the premise of human-caused global warming as an article of faith, and this takes it beyond being a topic of scientific analysis and has made it a moral imperative for those who have ridden in on a wave of enrapturement.

Remember that today we have a *research industry* of literally thousands of scientists, assistants, post-docs, and graduate students all completely accepting of the premise that: *humankind – and its use of fossil fuels and emission of greenhouse gases – causes global warming.* This unconfirmed but widely accepted supposition goes unquestioned into the mountain of publicly funded research that is in the billions of dollars per year range. Almost no one in this massive scientific undertaking questions the validity of the "input starting-point" for all this research: *that anthropogenic atmospheric greenhouse gases are the principal cause of a rising global surface temperature.* Remember, this is a conclusion that, while widely accepted as true, was never anything more than a rational supposition with only tenuous evidence to back it up. It depended on suggestive (or lowest level-of-certainty) evidence that was becoming more and more doubtful every day – just as the public perception of the notion – spurred on by enraptured activists – was seeming more and more certain every day to the general public.

"Denier"

Never has there been an advance in the field of science that didn't involve a denial of some aspect of the scientific thinking of the time. When it comes to climate change, today's use of the term "denier" is an anthropogenic global warming "movement" public relations effort to disparage those who disagree with the notion of human-caused global warming for legitimate scientific reason (the contrarians) by lumping them in with those who disagree because of scientific ignorance of the subject. Calling them all "deniers" is an attempt by the enraptured anthropogenists to belittle the legitimate science-based contrarians by equating them to the uninformed.

Chapter 7
Questioning the Unconfirmed Supposition

Until forever, Commandante.

— Che Guevara, 1965

THE RIO CLIMATE SUMMIT had been convened by the UNFCCC in 1992 to address the widely accepted "problem" of human-caused global warming[108]. Five years later, the Kyoto Protocol[109] in 1997 had been proposed, again by enthusiastic members of the UNFCCC with the hope of addressing the same problem. Both of these mass meetings of many of the world's government representatives who were already captivated, came well in advance of the "*proclamation of anthropogenic cause*" of January, 2001 that human activities were "without doubt" the principal cause of global warming (thus indicating the strong preconception bias of the times). The planners of these events were motivated by the powerful preconceived notion of human culpability. They saw no need to attempt to confirm what almost "everyone" already knew to be true. Thus, they thought it would be possible to establish firm greenhouse-gas emission policies worldwide based on the still scientifically unestablished but widely accepted idea among environmentalists and idealistic policymakers that anthropogenic greenhouse gases were the principal cause of global warming. However, at this point the noble concept of human beings taking responsibility ran headlong into the opposition of the highly skeptical and politically powerful U.S. industrial community and their congressional allies.

The Endangerment Finding

In 2007, the U.S. Supreme Court found that the Environmental Protection Agency (EPA) must determine "whether greenhouse gas emissions contributed to climate change"[110].

To provide the affirming answer they felt the court wanted, the bureaucracy at the EPA sorted through the thousands of papers of "climate research" that had already been completed to find those that showed the earth was indeed warming. As was pointed out earlier this was a considerable percentage of the $50 billion in research funded by the federal government over the previous several decades.

Next, they added the papers that considered the many adverse effects of global warming (or climate change): direct temperature effects; air quality effects; extreme weather events; disease and allergen effects; negative effects on food production, agriculture, and forestry; detrimental water resource effects; sea level rise; impaired energy and infrastructure effects; altered ecosystems; and adverse effects on wildlife. Thus, they built a seemingly compelling case for the EPA to control greenhouse gas emissions because the report was a massive compilation of global warming and climate change information, figures, statistics, and reports.

Notice, however, that this entire, massive collection of data was really about the effects of a globe that is *warming,* not one that is necessarily *warming because of human activities.*

The various reports' conclusions were virtually all based on the unconfirmed supposition that atmospheric greenhouse gases have a consequential effect on global temperature (the "*proclamation of anthropogenic cause*" which was actually a "journalistic notion" based on the BBC press report at the Shanghai environmental conference that scientists "were sure beyond doubt" that humans were responsible for global warming) – which was made back in January of 2001. The earth is indeed warming and these reports show projected effects of continued warming. The only thing missing was scientific evidence that human-caused emission of greenhouse gases such as CO_2 into the atmosphere was causing that warming (the determination of which was the objective of the JASON group that approached Jimmy Carter in the first place as well as the funding the congress had so generously offered). Rising quantities of greenhouse gases might actually have little to do with global surface temperature changes. Such a connection had not been scientifically established. Two things happening concomitantly does not necessarily mean one is causing the other. Bicycle thefts in Kansas City over time may show an almost exact correlation with plumbing fixture production in Wisconsin over the same time, but that does not mean there is a cause and effect relationship or even that the two are responding to the same impetus.

Perhaps knowledge of the growing number of believers – in addition to the sheer volume of paper being referred to – would supply the desired degree of certainty that the court wanted, because there certainly was no solid scientific confirmation coming from the literature search preceding the endangerment finding. Only the evidence that a warming earth had significant potential consequences could be provided because there was no evidence that humankind's emission of greenhouse gases into the atmosphere was consequentially causing that warming. But it didn't seem to matter that there was no evidence of humankind's consequential role in all this. It was a huge bundle of ominous information on a warming world that "everyone" believed, and that seemed to be what was wanted and what mattered.

After extensive study and review, the EPA study group found that, yes, the earth was warming and therefore carbon dioxide was a pollutant to be regulated[111]. *With this, the EPA and the courts – whether they realized it or not – had accepted the unconfirmed supposition that human-caused greenhouse gases were causing global warming without an iota of positive valid scientific evidence. All they had was confirmed evidence that the earth was warming, that the warming had severe consequences, and that millions of following members of the public believed that humans were responsible. The supreme court had completely abdicated their judicial responsibility to be the ultimate arbitrator of such matters in favor of a government bureaucracy whom they thought more qualified than themselves – but who were actually not at all qualified to make such a determination as demonstrated by their decision. They should have told the court they had no basis for such a decision and were not yet ready to make such a determination – but they too were believers.*

The endangerment finding now became the ruling that justified the many actions that the Obama administration took to curtail the emission of carbon dioxide into the atmosphere. The "Clean Power Plan" also called the "War on Coal" was the most far-reaching of these actions[112], the objective of which was to close down nearly all the nation's more than 500 coal power plants just as soon as possible.

A fortuitous coincidence was that the deviational drilling/horizontal fracturing (fracking) process came-of-age and rather suddenly became accepted by the industry. Also, newly-defined and better-defined natural gas deposits made huge quantities of low-cost natural gas available to replace the coal in power plants at somewhat lower CO_2 output (as well as lower cost) per unit of energy.

Prospecting

The arsenal of geological and geophysical prospecting tools for oil and gas has grown dramatically over the years. Traditional geologic and seismic prospecting technology has leapfrogged as advanced geo-acoustic imaging has been introduced[113]. *In addition, improved gravity survey, magnetic prospecting, bore-hole analysis, and geochemical prospecting techniques have all been introduced. These advanced techniques have become more and more able to detect and define deposits that were formerly undetectable or undefinable. This means the supply of gas and oil from formerly undefinable fields can now be exploited. Production will continue to grow as more and more "plays" of fossil fuels are discovered and characterized – although as some of them become harder to exploit, it can mean greater cost.*

As the twenty-first century progressed, many hundreds of scientists were engaged in climate change research that involved a highly questionable premise or unconfirmed supposition as the central conceptual basis for that research. Practically none of the scientists involved in this movement were working on the question: "Is this highly questionable premise or unconfirmed supposition valid?"

Even after the *proclamation of anthropogenic cause* in 2001, subsequent climate conferences such as the Copenhagen Summit of 2009 also went nowhere[114] – all for good reason. Too many key people felt that the premise that humankind's use of fossil fuels was the principal cause of global warming was a fallacious (or at least an as-yet undecided) notion. Despite the "proclamation" of the inner-core of climate scientists and the wide acceptance of that "proclamation" by the mainstream media combined with the forward momentum of the anthropogenic global warming movement, as stated before, there was simply no credible scientific basis for the anthropogenic premise. And the recalcitrant politically powerful contrarians, even if they did not completely understand the subtle details of the correct physics, knew when they were being taken for a ride by a propaganda campaign. In particular, there was considerable skepticism among the vested fossil fuel and peripheral industrial interests in the United States and their political allies.

But even the contrarians, try as they did, could not come up with a good alternate explanation of what was causing the earth to warm. This led to the attempt by some of the contrarians to dispute that the globe was even warming. Their clarion-call became "The earth is not warming", primarily because they had no explanation of what was causing the earth to warm... and that was an easy way out. There was so much misinformation and disinformation being bandied about that it was easy to surmise that some anthropogenic climate scientists had used questionable – if not outright fallacious data – which led to a "turnabout is fair play". Thus an attitude of: *We don't believe your data showing that the earth is warming.*

There was simply not enough known about what was causing the earth to warm to answer the question: "If it isn't human-caused atmospheric greenhouse gases causing the earth to warm, what is it?" The controversy was being won in the mind of the public by the anthropogenists because they had an answer, even if that answer had not been appropriately scientifically demonstrated (and ultimately was found to be false). The contrarians had to find evidence that the globe was not warming which put them in a bad position because the earth was indeed warming and virtually all the valid evidence showed that it was. Their anthropogenic adversaries seemed to have reason on their side because the contrarians had chosen the wrong battleground. If the question was "Is the earth warming?" then the anthropogenists were going to win. The anthropogenists had pinned their human-caused position on very flimsy evidence. And that evidence was beginning to deteriorate. Thus, they welcomed the shift from the fundamental question "Are humans causing the earth to warm?" to the new battleground "Is the earth warming?". This *turn* went a long way toward helping the anthropogenists win the first major battle. But the contrarians, still grumbling and with only their gut feelings to support their position, retreated, but did not give up, and they still had considerable legislative clout.

Without United States congressional endorsement, the anthropogenically-motivated international agreements to curtail greenhouse gas emissions would simply not have legs in the U.S. beyond what the anthropogenically-inclined administration could provide by executive order, or what could be interpreted to come under previously enacted legislation. Both of these routes were subject to future reversal.

With their *proclamation of anthropogenic cause* of January 2001, the inner-core had done a good job of reassuring the already-convinced – who didn't really need to be persuaded, but not such a good job of influencing anyone with a slightly skeptical "show-me" attitude who was looking a little deeper than the prevailing very simplistic logic path (elucidated earlier) for solid evidence of human-caused global warming.

But the proclamation did influence the journalistic community to accept the anthropogenic global warming premise. This turned them away from both the message of deteriorating anthropogenic evidence of global warming and the growing evidence of natural causes of global warming coming in patchy uncontextualized bits-and-pieces – not so much from contrarians, but mainly from findings about the sun and its fascinating multiple previously unrecognized or unappreciated interactions with the earth.

The Period of Equivocation

During the first decade of the twenty-first century no one really knew if the human use of fossil fuels was responsible for global warming or not. Yet it had been proclaimed that there was "no doubt" that they were and most of the general public seemed to believe it. The inner core apparently believed it was best for all if the "public" thought that was the case. And more and more people seemed to be accepting it without question. Those who did question were quickly dispatched as "deniers" and "everyone" seemed to accept that as well. So the conventional wisdom became that humans and greenhouse gases were the culprits.

If you are an expert, there are several problems with declaring that something is true – when you and a few insiders know that it may be true, or it may not be. The biggest problem might well be within yourself. If you repeat a fallacious statement enough times, and the reception is sympathetic and responsive, you might actually begin to believe it yourself. This clearly was happening within the inner core of climate experts.

But at the same time, new evidence was emerging that might answer the question: What is the real cause of global warming? It was becoming obvious to some scientists that changing atmospheric greenhouse gases were not powerful enough to cause the extent of global warming we were witnessing. Furthermore, evidence was emerging that the sun had a hither-to unrecognized slowly changing (multiple hundreds of years) cyclic magnetic field that was affecting earth's cloudiness which might affect the amount of solar heat reflected away from the earth.

But the inner core scientists directing the IPCC information agenda (the public-opinion guidance group) were extremely reluctant to relinquish their enrapturement with a concept they had extensively nurtured. They were not interested in even looking at competing potential causes. So they took no interest in these developments and simply continued to ignore them.

By now, the anthropogenic global warming "cause" seemingly had started to become more important than the scientific curiosity (and in a few cases, the moral integrity) that had always been so important in the field of science. The social philosophy more often found in political and religious movements of "*the only truth is in imposing the dogma*", which was already firmly entrenched in the environmental community, began to enter the anthropogenic climate scientific community. It was becoming increasingly difficult to question the newfound orthodox dogma. Some of these anthropogenic-oriented climate scientists had unfortunately transitioned from "impartial scientists" to "captivated observers" to "enraptured functionaries" to "activist ideologues".

The Paris Agreement of 2015 with nearly two hundred "believing" national delegations set carbon dioxide emission limits for various nations (including the U.S.) which the attendees from all nations agreed to[115].

The U.S. delegation thought they could impose limits on carbon dioxide emissions on American corporations by using the endangerment finding of 2009 that was supported by the courts. But congressional action had to be skirted because that was, at best, a tenuous path considering the skepticism and political dissent. Instead, the Obama administration started by trying to reduce the use of coal – the largest emitter of carbon per unit of energy. They used the many powers of the administration to leverage the direction of private industry. They disapproved applications for rail extensions for coal delivery to new power plants; for coal exploration and development permits on government-owned or controlled lands; and also simply denied applications for coal plants based on the fact that they emitted carbon dioxide[116].

They quickly succumbed to pressure from environmental activists to do the same with petroleum and disapproved pipeline applications[117].

In addition, "The Clean Power Plan", developed by the EPA, set limits on CO_2 emissions by state but allowed states on the same electric power grid to co-operate[118]. The hope was that virtually all coal-powered electric generating plants would be forced to close down. This seemed to be a viable possibility since cheap natural gas was coming on strong from the fracking of the abundant tight shale formations such as the Marcellus.

Natural gas became the new favorite of the environmental activist movement even though it too, as a fossil fuel, emitted copious quantities of carbon dioxide – although not as much as coal or petroleum liquids. It made for low-cost energy as well, so the drilling industry as well as the customers were happy. The only ones losing out were the coal industry and coal miners.

Some of these anti-coal executive decisions were quickly reversed when the Trump administration came to power. The Clean Power Plan began to be "studied" to see if it was necessary[119]. Virtually all this anti-carbon-dioxide activity was based on the unconfirmed supposition that increasing atmospheric greenhouse gases caused consequential global warming.

The newly elected president, Donald Trump, declared the Paris Agreement to be a "bad deal" for the United States, and said he would not honor it[120]. Initial reaction by an international community that had been led into the embrace of the concept of anthropogenic climate change by a coalition led intellectually, scientifically, and philosophically by the Europeans and Americans, (and financially, particularly by the U.S. government), was one of shocked dismay. This position signaled that the exceedingly generous international financial support of the whole concept of human-caused global warming by the United States was about to end. No wonder the many captivated international administrative and "receiving" participants were so dismayed. With no more United States supplied financial resources, many of the less developed nations would probably have to continue (or begin) to use the most economical method of supplying energy to their people, which was often the readily available, now disparaged, carbon-loaded, coal.

Since the Trump administration's original "position statement", the topic of global warming was rather quiet at the U.S. Government level until November of 2019 when they informed the UN that the United States would pull out of The Paris Accord at the soonest time allowed by the treaty, November of 2020.

The battleground within the United States had quietly shifted from the federal government level to the state and region levels and seemed to be settling in as a political party issue. What a shame! What is causing the earth to warm is not a political issue. Just because the "overwhelming majority" of scientists currently seem to believe in the general anthropogenic position does not mean that this perception will continue. *You must remember that the scientific inner core had never "found" that human use of fossil fuels was causing global warming. They had "found" that it was a possibility – and that since they were not close to determining whether increasing carbon dioxide was the cause of global warming or not, it was better if the public thought it was, so they would take the actions necessary to curtail greenhouse gas emissions into the atmosphere – just in case it later turned out to be the case.*

There is so much "new" scientific information just beginning to emerge, that once this information is absorbed by the greater scientific community, it will become abundantly clear to the vast majority of scientists that the systematically changing magnetic and radiational features within the sun and their effect on earth's sustained cloudiness (and its reflectivity of solar heat) as well as sun/earth proximity relationships are the important factors influencing the earth's temperature. In addition, (as you will soon see), the notion that human-caused greenhouse gases are the driving force behind global warming has already been determined to be fallacious.

Chapter 8
Is Carbon Dioxide Harmful?

Facts and science are hard masters.

— Kevin Sharer[121].

MOST OF THE CONCERN about the human use of fossil fuels falls into one of three categories:

- Mining and extracting practices that are sometimes detrimental to the people involved in the mining or drilling process or that create nearby environmental pollution, including visual, noise, and ash pollution.
- Air pollution, that is, the smoke particles, sulfur and nitrous oxides, aerosols, and harmful molecular and elemental byproducts of combusting fossil fuels (such as mercury or sulfur).
- The addition of carbon dioxide to the atmosphere, which – according to the anthropogenic global warming coalition narrative – causes global warming (and subsequent climate change).

The Environmental Protection Agency has been concerned with the first two of these factors since its formation in 1970 – and rightly so. Nothing should be done to minimize the appropriate legitimate efforts that this agency has done, is doing, or should be doing in the future to properly and legitimately protect our citizens from shoddy mining or extracting processes or to eliminate or reduce our valid and fair efforts to keep our air and water clean. In some process controls, however, the pendulum has swung too far, and overzealous testing and record-keeping requirements have been imposed that are based on standards that are questionable. Often there is more pain than gain in these excessive regulations. They are ripe for investigation, discussion, evaluation, and correction. Individuals with balance and perspective are more important than ideologues who mindlessly enforce unnecessary, overreaching requirements.

However, as the American physicist William Happer recently explained, carbon dioxide is not a "pollutant" but is an atmospheric enricher[122]. Plant life requires atmospheric CO_2 for its existence. The higher the atmospheric CO_2 content, the more plant life flourishes[123]. Controlled atmosphere studies of plant growth show that crop yields are enhanced in tests where atmospheric CO_2 content is higher than in the open air. Forests in the United States are more robust and vibrant as a result of increased atmospheric CO_2[124]. The current level of CO_2 in the atmosphere is very close to the lower limit of plant life existence[125], but fortunately it is increasing both as a result of human-caused emissions from fossil fuels and as a result of natural processes such as volcanos and off-gassing from a naturally warming ocean.

Yet this increased atmospheric CO_2 level is not detrimental to animal or human life. All animals exhale carbon dioxide with every breath. We animals need oxygen to survive. While atmospheric carbon dioxide is not needed for animal existence (except for the plants we and our livestock eat), it is not detrimental until it reaches very high levels, which, incidentally is far higher than would be reached in our atmosphere if the entire world supply of recoverable fossil fuels was burned up and the resulting CO_2 was emitted into the atmosphere. The atmospheric carbon dioxide matrix (graphic 8.1) shows some of the key points of interest in atmospheric carbon dioxide content in the atmosphere.

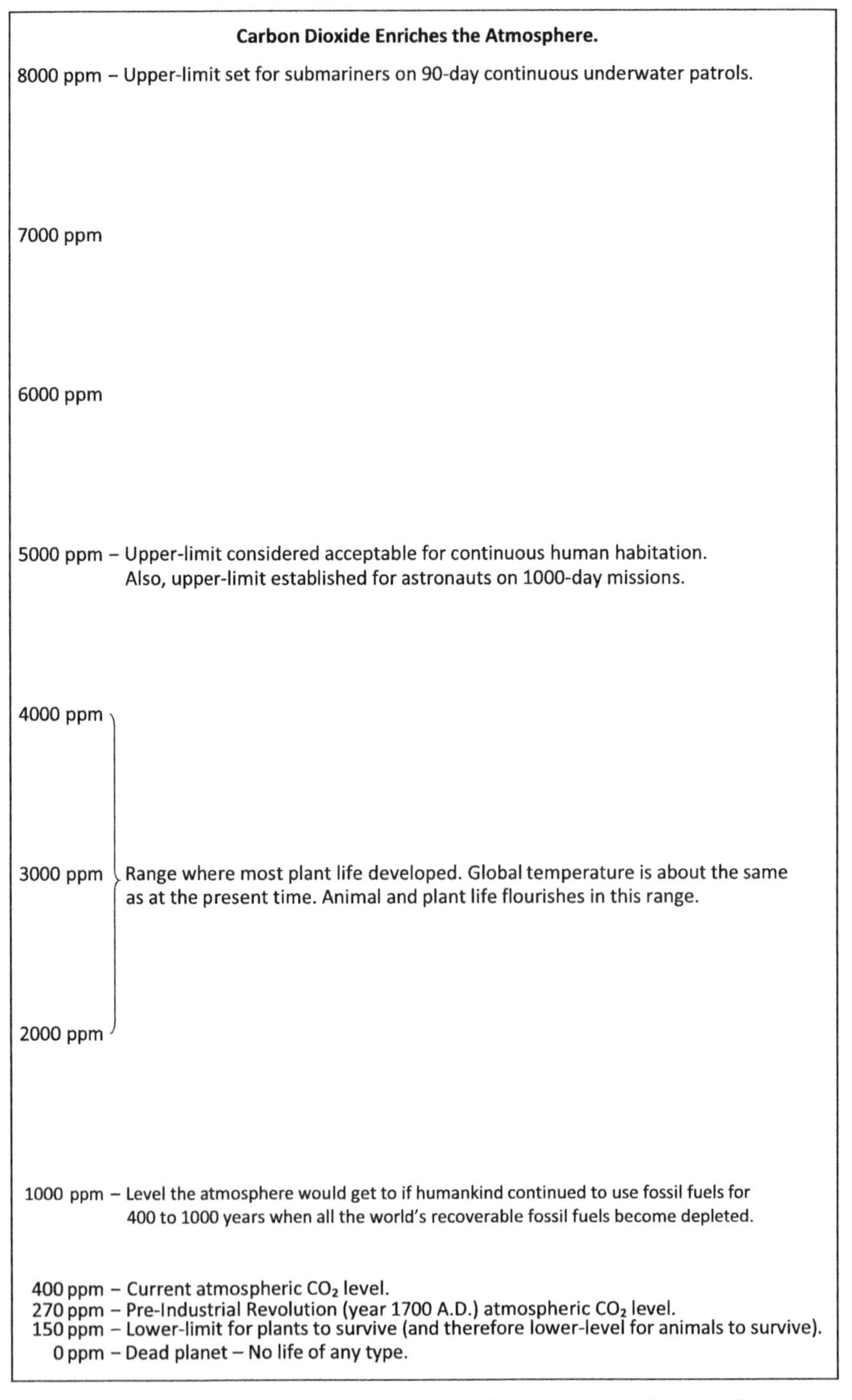

Graphic 8.1 **Carbon Dioxide Is Not an Atmospheric Pollutant.** W. Happer/ E. Pryor [126].

So the prime case for carbon dioxide being a "pollutant" rested on the widely accepted but scientifically unconfirmed supposition that its increase in earth's atmosphere will cause consequential detrimental global warming. As time has passed, it has become increasingly clear that this unconfirmed supposition is invalid, as will soon become evident. (Ocean acidification, another "pollution" factor that is sometimes suggested, is discussed later, in appendix D.).

Extinctions

The causes of various animal extinctions (one of which is the dinosaur extinction of 64 million years ago) make for interesting speculation. An asteroid impact causing an "impact winter" or possibly a moving plume of magma in the mantle underlying the earth's crust evidenced by the Deccan Traps in India were two such postulations for that particular extinction, (although the impact theory now seems to have won out). Other mass-extinction theories have been proposed for various times in the paleo (distant) past. But in addition, it definitely appears that various species of animal-life appear and then naturally disappear after a significant long-term interval in the hundreds of thousands to millions of years range. These are called "background extinctions" – which just seem to happen to individual species for reasons applicable to that particular species.

One speculative possibility that should be added to the list of possible causes for mass- extinction is a progressive lowering of earth's atmospheric carbon dioxide content wherein plant life becomes so dominant that this kingdom begins to "use-up" the small supply of CO_2 in the atmosphere at a faster pace than it is being replaced. Various natural sources of atmospheric CO_2 contribution to the atmosphere include: animal exhaling, volcanic activity, off-gassing from a warming ocean, and creation (or transmogrification) of carbon atoms from nitrogen atoms in the atmosphere by bombardment of cosmic radiation from outer space and from ionic bombardment from the sun. Specific plants that required more carbon dioxide than others would die out first (along with the animals that could only live on those types of plants) – thus causing selective plant and animal extinctions. There is paleo ice-core evidence that at several points in the distant past, the earth's atmospheric carbon dioxide content got all the way down to between 170 and 180 ppm[127], which is not far above the 150 ppm, considered to be the minimum point for plant life existence. Thus, it is easy to speculate that insufficient CO_2 is a potential cause for various mass extinctions – both past and future.

Another, not frequently mentioned possible cause of extinction is the simple global cooling caused by the separation of the earth and the sun as a result of orbital Milankovitch cycles perhaps exacerbated by the gravitational pull of "rogue" extra-galactic bodies that randomly enter the solar system. The severe cooling found in these temperature excursions can easily explain extinctions among various populations – both animal and plant – that can't either adapt or move. One such event may have occurred about a half million years ago when the Vostok ice core record suggests a change in the pattern of the earth's naturally changing temperature in accordance with the current 100,000-year Milankovitch cycle.

Is the Consensus Always Right?

The wide acceptance of the conclusion that humans and their emission of greenhouse gases into the atmosphere is responsible for global warming is not the first time nor will it probably be the last – that large groups of people, both scientists and the general public, have been led down a path by an alluring scientific-sounding concept. Over the years there have been a number of such suppositions (sometimes ideologically reinforced) that became the popular consensus – and then spread until nearly everyone believed – until the supposition finally was found to be invalid. The following are examples of where acceptance of a scientific concept by the overwhelming majority of scientists was good enough for a long enough time to convince the overwhelming majority of the general public – but later that widely-accepted premise proved to be incorrect:

The earth is flat.

This was believed to be true until humans began to explore far enough to notice the curvature of the earth. The only place that this was visually evident was on a relatively calm ocean horizon as another ship or perhaps an island began to disappear from the bottom up or began to appear from the top down.

The sun rotates around the earth.

Ancient people could "see" that the sun was rotating around the earth, as it came-up every day, moved across the sky, and finally sank below the horizon. After considerable deliberation, scientists of the time "established" that the earth was the center of the universe, and the stars as well as the sun – also rotated around the earth. This scientific consensus was termed "geocentric" and was fostered by Ptolemy (and others) in the second century AD. It lasted for many centuries until Copernicus, in the mid-1500s, proposed the radical concept that the earth and planets rotated around the sun[128]. (See further discussion in chapter 17).

The elements are: earth, air, fire, water – and aether.

In ancient Greece, it was believed these were the four elements, plus the "void" of space[129]. The theory worked well for a time until there appeared to be more elements[130]. As time has passed the number of elements has dramatically increased. We have now slowly gone up from four to well over one hundred elements – with perhaps more to be discovered.

Human sacrifice is necessary to prevent natural disasters.

Human sacrifice was an important ritual in the Western Hemisphere among the native peoples – particularly among the societies in Mesoamerica such as the Maya and the Mexica who believed that human sacrifice nourished the gods[131]. At the time of European contact in the 1500s, it was estimated that more than five thousand (and perhaps many more) people were sacrificed each year by the native peoples. The preferred method of sacrifice was for four priests to hold the candidate on a ceremonial stone table at the top of a flat-topped stone pyramid while the chief priest, using a ritualistic knife with obsidian blade, entered the body through the abdomen of the squirming candidate, sliced up through the diaphragm, and extracted the still-beating heart which he held high as it continued to beat – usually to roars of approval from the crowd below.

It was rumored that one group of Native Americans, for example, thought that if a person was not sacrificed each day the sun would not come up the next day. Accordingly, they always sacrificed at least one person each day (a captured enemy or clan prisoner, ideally, but if none were available, anyone would do), and, sure enough, the sun always rose the next day, thus confirming their belief.

The last confirmed ritualistic human sacrifice event to have occurred in the western hemisphere was at the time of the massive earthquake (9.6 on the Richter scale) in 1960 in southern Chile (which sent out a tsunami powerful enough to kill dozens of people in Hilo, Hawaii). This occurred in the modern era following World-War II[132]. The Mapuche people of southern Chile sacrificed a five-year-old boy to stop the earth from trembling. The boy's mother was a domestic servant in Santiago, many miles distant and the boy had been left in the care of relatives who offered him up to the tribal elders for the honor of being sacrificed. This was felt by the elders to be essential to stop the shaking earth (and which it apparently did after a few more days following the sacrifice, confirming the belief that such an action had been necessary). An exceedingly wise, well-educated, then-retired Chilean supreme court chief justice was dispatched to determine if any charges of wrongdoing by the elders (or others) would be appropriate. He determined that they had "*acted without free will, driven by an irresistible natural force of ancestral tradition*" and should not be charged[133]. (This is an example of an entrenched, emotion-based – or enraptured – belief, not unlike that of what the notion of anthropogenic global warming is becoming.)

Bloodletting is a cure for most human ailments.

Bloodletting was widely practiced throughout the civilized world for hundreds of years. Often it was the only medical option available to the then primitive medical practitioners (perhaps other than a mustard-plaster to the solar plexus). Both King Charles II of England (1685) and more than a century later, George Washington (1799), suffered minor ailments that were treated by bloodletting and then died. Thus indicating that bloodletting didn't always work. On the other hand, a French sergeant in 1824 was bayonetted in combat and fainted, presumably due to loss of blood. He subsequently had a series of bloodletting sessions until a total of 13 pints of blood were removed in addition to the original stabbing loss. Ultimately, he recovered – thus confirming the efficacy of the practice of bloodletting[134].

Heavy objects fall faster than light objects.

Sometimes after a law of physics is "declared", it is believed for many years before being successfully challenged. Let's look at Aristotle's law of gravity which stated that heavy things fall faster than light things – based on his comparative experiments of simultaneously dropping a metal bar and a feather (the concept of "air" and its possible effect of slowing down the feather had not yet been conceptualized). This law of gravity was proclaimed by Aristotle around 350 BC. Thus the concept that heavy things fall faster satisfied everyone for nearly two thousand years. Then de Groot in the Netherlands and soon thereafter (and more famously) Galileo challenged this law in the 1500s. Galileo performed his dramatic dropping of a wooden ball and a lead ball from the Leaning Tower of Pisa and, whoa – both landed at the same time thus ending the reign of Aristotle's law of gravity. Newton's laws then took over until the twentieth century.

Stress is the cause of duodenal ulcers.

Until as recently as the late twentieth century, duodenal ulcers were thought to be caused by a combination of stomach acid and stress. Four famous over-the-counter pharmaceuticals were independently and expensively developed to cure this scourge. Then an Australian physician discovered that a bacterium – helicobacter pylori, was the true cause and ulcers were treatable with antibiotics. The four expensively developed "acid-controllers" were quickly diverted for use in controlling GERD and acid indigestion by millions of people who apparently still needed relief from these more minor maladies and who happily succumbed to the very effective advertising for these over-the-counter remedies.

The phlogiston theory becomes accepted by the scientific community.

In the fifteenth and sixteenth centuries, it was thought that phlogiston was a substance – sort of the opposite of oxygen. For example, the phlogiston theory held that a flammable material "dephlogistonated" when burned. Likewise, the act of breathing presumably expelled phlogiston from the body. This theory, although controversial, was the overwhelming conventional scientific consensus for more than 100 years from the 1670s to the 1780s before being superseded by modern science[135] (which is generally attributed to having begun in concert with the Industrial Revolution).

All these (and many other) concepts showed, at least for a while, some apparently compelling justification for being correct, spread throughout the scientific community, and then the general public (perhaps picking up some ideological momentum along the way). All were a part of human beings' continuing search to understand their natural surroundings. Is the concept of consequential anthropogenic global warming just another step in the continuation of that search? Is it perhaps just another, but more modern, phlogiston theory?

Early in the twenty-first century, many members of the peripheral scientific community who had not yet come under the anthropogenic global warming spell, proceeded on the path of scientific exploration (as scientists are expected to do) – by looking further into the mysteries of materials and processes – some of which turned out to relate to global warming. In particular, the examination and characterization of the processes involved in the interior workings of the sun were becoming more and more revealing. These scientists were going where the scientific exploration took them, and humankind's understanding of the science moved forward, as it always does, despite statements by those enraptured by the anthropogenic global warming notion that the "case is closed", the "science is settled" and the "science doesn't change"[136].

A number of independent scientists were uncovering more and more very interesting findings: things that even went so far as to raise questions about the validity of the premise that the peripheral or "downstream" scientists were using as the starting point for their research – the "proclaimed" but as-yet unsubstantiated supposition that increasing carbon dioxide content in the atmosphere was the principal cause of global warming.

Chapter 9

The “Case” Begins to Deteriorate.

How dreadful knowledge of the truth can be.

— Sophocles, 429 BC

THE ENVIRONMENTAL PROTECTION AGENCY summarizers were busy building a case to affirm that humans were indeed responsible for global warming and that the human emission of greenhouse gases should thus be regulated. In addition, the general public as well as much of the political establishment in the United States and around the world was becoming attracted by the premise of the anthropogenic global warming “movement” (which was steadily growing). At the same time that this anthropogenic global warming “message” was spreading and the movement was growing more and more appealing to larger groups of people throughout the world (it all sounded so logical) – widening cracks began to appear in the evidence supporting the anthropogenic premise. These were cracks that members of the tightly knit anthropogenic global warming inner core were very reluctant to explore.

The physics-based evidence for human-caused global warming began to be examined at a more basic level by scientists who were from outside this tightly-knit group. The anthropogenic inner core had used the Kiehl and Trenberth paper of 1997 as the theoretical basis for their conclusion that atmospheric carbon dioxide had a consequential influence on global temperature[137]. And now the interpretation of the physics of that paper was being reconsidered by outsiders. It appeared to some of the molecular spectroscopic community that splitting the infrared absorption effect 50 percent to each in the zone overlapped by both water vapor and CO_2, gave far too much of the greenhouse effect to CO_2 and far too little to water vapor[138].

Since this is one of the basic reasons that the concept became so widely accepted in the scientific community (and ultimately by the general public), we give a more expanded explanation of the physics in appendix A. Here we will say there appeared to be an unfortunate crucial misinterpretation of the physics that dramatically overexaggerated the influence of changing atmospheric CO_2 on global surface temperature. Specifically, the interpretation difference was rooted in the decision by climate scientists to apply 50 percent of the heat absorbing effect of outgoing earth heat radiation to each water vapor and CO_2 in the molecular spectroscopic zone where their infrared absorption bands overlap[139]. *If we removed all the CO_2 from that overlapped zone, would there be any effect on global temperature? No, because water vapor (at least 12 to 15 times the concentration of CO_2) would still be absorbing all the outgoing heat in that overlapped zone. Likewise, if we added double the amount of CO_2 in that zone would there be any effect on global temperature? Again, no, because water vapor is already absorbing all the outgoing heat in that zone. So in that overlapped zone, 100 percent of the greenhouse effect must go to water vapor when we are trying to determine the temperature effect of a change in CO_2 content in the atmosphere.* Although the 50-50 split between water vapor and

CO_2 "sort of looks" fair, it is not. It must be 100% water vapor as you can see from the above[140]. This is the central basis for the conclusion that humankind's emission of greenhouse gases into the atmosphere is the principal cause of global warming and climate change and thus is the heart of the anthropogenic global warming problem. (See appendix A for a more detailed explanation.)

Secondly, Kiehl and Trenberth in their 1997 paper used the molecular spectroscopic configuration of water vapor and CO_2 from the HITRAN database published in an encyclopedia in 1992, which was the only one available to them at that time. But the water vapor data from that database were preliminary and were later revised and updated[141]. The "window" in the infrared absorption band of water vapor became more than 25 percent smaller during the early years of the twenty first century (while the CO_2 "plug" remained the same). This meant that a much-increased amount of the two absorption bands became overlapped[142] and the effect of CO_2 became further diminished.

These two very important – but apparently unintentional – flaws have been carried through years of scientific discourse within the climate science community ever since.

In addition, the feedback effect of increasing atmospheric water vapor was examined by neutral scientists and then seriously questioned. The inner core had estimated this figure to be enough to amplify a calculated 1°C rise in global surface temperature into a 3°C total rise in global surface temperature when atmospheric CO_2 was doubled. Various studies indicated that this water vapor feedback amplification factor was much too high[143].

Appendix A shows where the water vapor and CO_2 absorption bands overlap. The CO_2 absorption is redundant because the water vapor is already absorbing the earth's outgoing infrared heat in that overlapped zone. Thus, in that zone, some, (or in fact any) amount of carbon dioxide has almost no effect on global temperature because of the near-universal presence of water vapor in our atmosphere. There is another small zone where the two do not overlap, and in that zone doubling the presence of carbon dioxide would have a small (or trivial) effect on global temperature.

To get the appropriate full picture we must expand and jump ahead for a moment in order to understand the seriousness of these unintended scientific flaws and misinterpretations:

Climate Sensitivity

Scientists attempting to quantify the amount of temperature rise that would result from a doubling of atmospheric carbon dioxide were finding a multiplicity of results. Real world empirical research was showing fractions of a degree Celsius ranging from: 0.4°C [144] to 0.45°C[145] to 0.5°C [146] to 0.7°C[147]; while the mainstream anthropogenists who were accepting the 50/50 split in forcing between water vapor and carbon dioxide were all getting much higher results between about 2°C and 4.5°C with the IPCC's "best estimate" of 3°C[148]. With this kind of near order-of-magnitude difference between "camps", it was becoming obvious that the mainstream anthropogenic oriented scientists were apparently unaware of the "overlap" problem which, if they knew about it and accepted it, would invalidate their main thesis that human-caused greenhouse gases were responsible for global warming. And they apparently were not willing to even investigate anything like that. And why should they? "Everyone" seemed to be jumping onto the anthropogenic bandwagon.

In 2014, the relatively new-to-the-scene Dr. Hermann Harde from the Helmut Schmidt University in Hamburg, Germany looking in detail at the physics from the molecular spectroscopic level, was able to come-up with a very comprehensive, more realistic fully-quantified and properly proportioned allocation between atmospheric water vapor and carbon dioxide[149]. This resulted in a 0.6°C rise in global surface temperature for a doubling of atmospheric CO_2 (a trivial figure). This climate sensitivity number includes the corrections for both the absorption band overlap (as revised by the updated HITRAN database) and the corrected water vapor feedback. Thus this 0.6°C figure is the updated trivial amount to be compared with the inner core's clearly unfounded obsolete five-times-higher, consequential 3°C figure, the "working" figure so widely accepted among peripheral global warming scientific researchers (and that is the basis for alarm by so many in the environmental, journalistic, and political communities).

But by the time this correct 0.6°C figure was revealed in proper quantified form in 2014, the concept of the dominance of human-caused global warming had become so cemented into the entire secondary (peripheral) climate-science community (and had received such sweeping public acceptance), there could not be an endorsement or even an acknowledgement of this new lower figure by either the IPCC or the inner core. In fact, this finding that the global surface temperature effect of doubling atmospheric CO_2 was much less than that claimed by the inner core was just not palatable to a climate science-community that had become so comfortable with the wide acceptance by the public of their version of the "science" that they could see no reason to even look into any of this "new thinking". Thus, it was simply ignored by both the inner core and the IPCC. And since they remained the only "establishment" spokespeople on this topic available to the journalistic community and the public, it still is not widely known or understood.

The "anthropogenic concept" simply had gone so far down the wrong fork in the road that it was going to be very difficult to get back on track at this point. This is what can happen when a deadline forces a proclamation of fact by the experts in a field where there is considerable preconceived belief and yet the science is still unfolding. The inner core of experts had committed themselves at the "*proclamation of anthropogenic cause*" of January 2001 and there was no provision for revision. They now had to become the defenders of the proclamation rather than continue to explore the science. It might even be said that for these now unfortunate "enraptured" scientists, the search for scientific truth had been overtaken by the "movement" – and its demand for acquiescence to and acceptance of the prevailing conventional wisdom (the "*adhesion becomes a duty*" of Schopenhauer.)

The Scientific Misconception

The conclusion that a doubling of atmospheric CO_2 would result in a 3°C rise in global surface temperature had already spread through the peripheral and general scientific communities and was being used by them as a fundamental-truth. "Climate science" for these legions of scientists became the study of the environmental and human effects of a change in climate caused by human activity which they had been led to believe was predictable based on the estimated use-rate of fossil fuels by humans – and that global temperature would rise by three degrees Celsius if atmospheric CO_2 levels doubled, based on that use. They did not think it was their task to look into the assumed basis for this conclusion – that was now a "given", or a starting point. Thus, all sorts of grim environmental problems were being predicted.

As this onslaught of pictures-of-environmental-disaster "snowballed", the public became more and more alarmed. Environmental activists called for political actions to "stop the madness". Yet the conclusion that a doubling of atmospheric CO_2 would result in a "consequential" rise in global temperature had not been scientifically confirmed and was actually an invalid concept, although that was not (at that time) understood by the anthropogenists. Unfortunately, if the supposition that there was a consequential effect caused by a doubling of atmospheric carbon dioxide was admitted to be fallacious, all the billions of dollars of research conducted in the last few decades and all the conclusions drawn would be of no value because they were based on a false assumption.

As pointed out earlier, the authors of the key theoretical paper upon which the whole scenario of human culpability was based thought the results of their paper were: "clearly not exact..." and there were "uncertainties and issues" remaining[150] and "this was just one scenario". On the other hand, the press began to state that the concept of anthropogenic global warming was believed and endorsed by "over 90 percent" of the climate science community or, alternatively, "the overwhelming majority of scientists". The enraptured environmentalists along with the journalistic community, picked-up on the wordplay and promoted the new catchphrase: "The established science of climate change".

So, at that point in the early twenty-first century, although it was not widely known, the physics basis for a consequential effect on global temperature by increasing atmospheric carbon dioxide became recognized by at least a few climate scientists as almost certainly invalid. But many other climate scientists were not even aware of this disturbing development, for it was not publicly discussed by the inner core or the IPCC who were the principal "mouthpieces" of climate science. There were some in the inner core who did realize that their theoretical postulation apparently had become invalidated or at least was being seriously questioned. But they were "believers" – and they were not about to give up. They had to become very creative to try to support their preconceived notion. Now they had to come up with some other method of showing that their conclusion that humankind was responsible for consequential global warming was still valid, because it certainly looked like their physics evidence had already been conclusively invalidated.

So these members of the inner core quietly switched their emphasis from the theoretical evidence to the observational evidence. There was not a hint of deviation from the path of human-caused global warming, there was simply a shift in the question: "*how are we going to continue to try to convince the remainder of the scientific community and the public to accept our premise, which was now lacking a valid physics basis?*"

At that point, with the theoretical evidence being questioned, the hockey stick (graphic 3.1) and the Vostok ice-core data with matching global surface temperature and atmospheric CO_2 versus time (graphic 3.2), now became the main supporting evidence – with the supercomputer climate models as backup.

Methane and Nitrous Oxide

Before we proceed, it may help to understand why atmospheric changes in powerful greenhouse gases such as methane and nitrous oxide have essentially no influence on global temperature or climate. While atmospheric increases in carbon dioxide are primarily caused by the human use of fossil fuels, atmospheric increases in methane and nitrous oxide are primarily caused by agriculture[151]. Methane is also the prime component of natural gas so "leaks" in the system of natural gas extraction and distribution are a concern of anthropogenic concerned activists. Methane and nitrous oxide are often cited for their potentially harmful effects on global temperature usually either by environmentalists or scientists working in other arenas who have been swept along by the anthropogenic movement. Some of these anthropogenic "worriers" fear that as the earth warms, for whatever reason, permafrost – which contains huge amounts of methane – will melt and release more methane, which will then initiate a feedback loop of more global warming, more permafrost melting, and more methane release[152]. This would be a valid concern if we did not have abundant water vapor in our atmosphere.

Methane and any increases in methane content in the atmosphere have practically no effect on global temperature because of the configuration of methane's infrared absorption band as it relates to water vapor's band. Methane's band is essentially completely overlapped by water vapor's in the limited zone affected by earth's outgoing infrared heat. Thus, for the same reason that applies to the partial overlapping of water vapor's infrared absorption band with carbon dioxide's, the earth's outgoing heat is already absorbed (in that zone) by water vapor over virtually the entire band occupied by methane[153]. So, in the earth's atmosphere, methane has virtually no effect (just as CO_2 has only a trivial effect) on global temperature.

The reason for the lack of clear understanding on this subject is that scientists tend to look at these gases "in isolation". They are entirely correct in their pronouncements of the greenhouse effect of these gases, but it is a theoretical finding that would be valid only if the gas was present in the atmosphere "in isolation" (without water vapor also present) – which it is not here in the earth's nearly completely water vapor saturated lower atmosphere. One of the problems here is the water vapor band was preliminary in the 1980's, and this preliminary (tentative) band was used by scientists looking at this arena. A rendition of this band published in a scientific encyclopedia in 1992 showed a wider "window" in this band that was later "closed in" when the HITRAN database was updated in the early part of the twenty first century. This "closing in" of the band caused more of an overlap between the water vapor and methane bands – thus slightly changing the methane effect on global temperature. (See graphic 9.1)

A similar effect is true for nitrous oxide whose infrared absorption bands are also overlapped by water vapor's. (Again, see graphic 9.1).

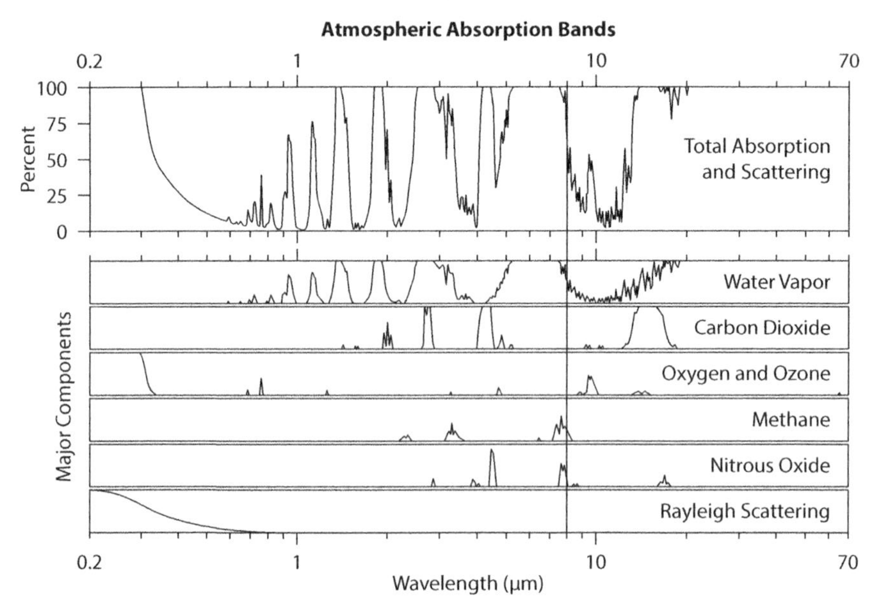

Graphic 9.1: **Atmospheric Infrared Absorption Bands.** Spectroscopic drawing. R.A. Rhode[154].

In graphic 9.1, one can see that both nitrous oxide and methane have an absorption "blip" at between 7 and 8 microns (μm), which is within the range of outgoing infrared heat for earth (The other "blips" are outside that range and thus have little to no effect on earth's outgoing heat). However, the water vapor absorption band is also present in that zone. Since water vapor will be absorbing the outgoing heat whether or not there is any methane or nitrous oxide present, changes in the amount of methane or nitrous oxide in the atmosphere will have very little effect on global temperature. Climate scientists who do not look at the updated full field of the spectroscopy (and apparently most inner core climate scientists and virtually all peripheral and downstream scientists were either looking at earlier versions of the HITRAN database or were not looking at the spectroscopy at all), miss a very important aspect of the influence of greenhouse gases on global temperature. They miss that water vapor – which is nearly always present in that same absorption band zone – will do the job, and that the trace greenhouse gases methane and nitrous oxide are in-effect, redundant, and they have little to no effect on global temperature[155].

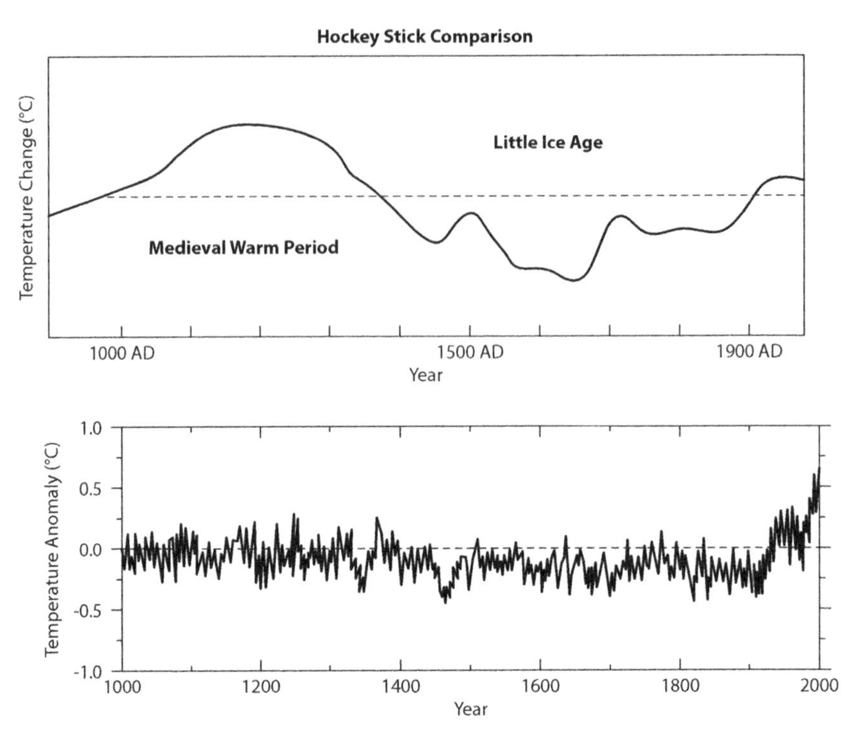

Graphic 9.2: **Hockey Stick Comparison.** Comparative graphs: Upper, IPCC First Assessment Report. (1990); Lower, M. Mann, IPCC Third Assessment Report (2001)[156]. [Time →]

- The upper graph in graphic 9.2 comes from data in the original 1990 IPCC First Assessment Report. It depicts global surface temperature for the last 1000 years with the Medieval Warm Period and the Little Ice Age intact (but with no quantification of temperature because of partially unsettled proxy data quantification).
- The lower graph is based on data from the 2001 IPCC Third Assessment Report, and is quite different from the earlier graph.

The hockey-stick evidence began to disintegrate with the revelation that the 1000-year temperature record displayed in the 1990

IPCC First Assessment Report had been altered a decade later in the 2001 IPCC Third Assessment Report. In the First Report the higher global surface temperatures of the well-established Medieval Warm Period were shown (with temperatures probably the same as or slightly higher than now). This rise peaked around the 1200s and its slowly diminishing effects lasted to nearly the fifteenth century during which time the Vikings populated a "greening" Greenland, were primarily farmers, and raised livestock as though they were still back in southern Scandinavia. The lower temperatures of the Little Ice Age from about the late 1600s to the early 1700s (when the Vikings left Greenland as it began to ice over again) were also shown. All of this was well before the Industrial Revolution and surging human use of fossil fuels.

In the "altered" 1000-year temperature flat-handle hockey stick curve shown ten years later by Sir John at the Shanghai Environmental Conference and also in the IPCC Third Assessment Report, the temperature was depicted as being much lower during the medieval warm period and essentially much more level during these times. And it did not begin to rise until the modern era – to roughly match the rise of anthropogenic atmospheric CO_2.

To further complicate matters, the East Anglia climate email release (popularly called Climategate) revealed an email that lamented "We have to get rid of the Medieval Warm Period." And lo-and-behold shortly thereafter the new "altered" hockey graph stick magically appeared from a helpful recipient of the lamenting email. It showed the more-or-less level global surface temperature during what had been an apparently meaningful, if not precisely quantified temperature-rise during the Medieval Warm Period; and what had been a subsequent consequential downturn during the Little Ice Age on the first graph, was also smoothed and flattened. Resourceful alterations to graphs like these can help validate a point-of-view, but they can also initiate a firestorm of controversy.

This brings us to the polemical topic of Climategate.

Chapter 10
Climategate

If you want a friend, get a dog.

— Albert J. Dunlap c. 1990

In 2009 more than 1000 emails and data files from the East Anglia (UK) climate research unit (CRU) were leaked by an anonymous whistleblower. These documents revealed some very disturbing practices in the way the inner core of climate scientists conducted their reporting of the facts and how their conclusions were drawn. After the emails were analyzed by pertinent scholars, review panels at various climate research facilities, particularly at East Anglia (UK) University and at Penn State University (U.S.), were convened to determine if any punitive action against climate scientists at their institutions was warranted. As well, several political investigative panels in the UK were convened, including one by the House of Commons. The controversy was quite dichotomous with anthropogenic climate critics (the contrarians or deniers) declaring this disclosure to be the most significant scientific scandal ever uncovered, while supporters of the anthropogenic-position declared it a non-event and claimed that nobody did anything wrong except to "purloin" some private emails. The main emphasis of the allegations of wrongdoing by the anthropogenists were on the alleged falsification of temperature records; the "loss" of such records; and on pressuring scientific-journal editors to not print scientific papers that supported the contrarian position by refusing to peer-review contrary papers submitted for publication.

In general, the internal investigatory committees tended to be sympathetic toward their accused colleagues and protective of the reputation of their institutions as well as extremely defensive of the human-caused aspect of global warming and the integrity of all IPCC findings. During the inquiries, the questioners stuck strictly to the accusations initiated by the "purloined" emails and did not dig any deeper.

Collective findings of all the various investigative panels were that an inner core of like-minded, anthropogenic-oriented climate scientists had engaged in routine and systematic:

- data manipulation.
- misapplication or failure to apply correction factors to selected data sets.
- suppression of criticism.
- exaggerated rhetoric.
- an attitude so focused on "winning" that the degree of certitude of the case for anthropogenic global warming was greatly overstated.
- an agreement to refuse to peer review contrary scientific papers thus preventing their publication in prestigious scientific journals.
- and finally, it was found that the peer review process, which works so well in most of science, does not work well at all in climate science.

All of which, while falling short of "gross scientific misconduct" constituted serious deviation, modification, or misuse of data (and prompted various letters of disapproval and several demotions of position – but no lasting severe punishment of the perpetrators)[157].

Another significant finding was that the appropriate caveats and expressions of uncertainty were often in the original research work but that these caveats were stripped away when the findings were presented by secondary users such as the IPCC summarizers, journalists, or various government agencies[158]. And seldom, if ever, was this publicly pointed out by the applicable scientists who were apparently perfectly happy to let the misconception of greater certainty stand. The question "Who should have stepped in to restore this qualifying data so the public would understand the limitations of the research?" was apparently never asked.

The anthropogenic-global-warming-friendly press pointed out that – since no one was severely punished – no serious wrong was done. Accordingly, now most of the general public has no idea what Climategate was all about or they think it was an innocuous bump in the road that was brought on by some of those evil climate deniers who were in concert with the devil himself.

Sometimes it is possible to present valid but potentially troubling information in a more favorable light, by "adjusting" the parameters of a graph without modifying any data sets. Simply not including data in the graph that would tend to conflict with your position or, even worse, withdrawing such data is definitely inappropriate. One's best bet would be to try to establish that the old data were somehow contaminated. Altering data-sets that conflict with a desired outcome is an absolute no-no. Switching proxies that could be "better" (or that conform better to a favored conclusion) is another highly questionable avenue.

The investigators found no actual alteration of a data-set, which was the only action considered serious enough for severe punishment – so there were no dismissals per se. There were departmental transfers and reductions in responsibility, however. Since the questioners never got into the sticky gray areas of this controversy, the public (and the questioners) never got a good picture of what really happened. Overall, the public awareness section of the IPCC, did a first-rate job of "whitewashing" Climategate and it soon disappeared from the public radar screen.

The Hockey Stick Controversy

In the IPCC Third Assessment Report of 2001, the hockey-stick had been altered from its original configuration in the First Assessment Report of 1990. In IPCC-1 (FAR), it showed significant temperature swings during the last millennium, with a Medieval Warm Period (peak from 950 to 1250 A.D.) with temperatures as high or even higher than now as well as a lower temperature during the Little Ice Age. In IPCC-3 (TAR), the temperatures during these two periods were much flatter. There is confirming geographic evidence that both of these periods did exist, despite climate scientists' inability to arrive at a consensus-quantified set of figures for just how hot the warmest temperatures of the Medieval Warm Period were.

Graphic 10.1. Historical photograph. **Remains of Viking Church, Hvalsey, Greenland[159].** Hundreds of abandoned small settlements and dozens of churches have been discovered along the fjords of southwest Greenland – such as the one shown here photographed by Frederik Carl Peter Ruttel (1859- 1915).

So there are widely documented observations (see graphic 10.1) that the Vikings had moved to Greenland and became farmers and husbanded livestock because it had become warm enough to do so during the (subsequently unacknowledged) Medieval Warm Period and then 300 to 400 years later these Viking farmers had moved back to mainland Scandinavia because it became too cold (during the subsequently unacknowledged Little Ice Age). There is substantial additional geographic evidence throughout Europe that these two climate periods did occur. Warm weather grape plants from the south of France thrived far north of their "home" when transplanted to Scotland during the Medieval Warm Period. Extended "freeze-overs" of the Thames River in London occurred during the Little Ice Age (see graphic 10.2).

Graphic 10.2. Historical Oil Painting by Abraham Hondius: **"A Frost Fair on the Thames at Temple Stairs" (1684)** ©Museum of London[160].

The painting shown in graphic 10.2 depicts a festive celebration held on the solidly frozen Thames River in London. These were held on numerous winters during the Little Ice Age – which did not happen according to the anthropogenic narrative as shown by the flat-handled hockey stick. Somewhat later, after the existence of the Little Ice Age became quite evident from both geographic and instrument (thermometer) records, the hockey stick was "adjusted" a second time by inner core scientists to show cooler temperatures during the Little Ice Age. But by then the damage had been done to the hockey stick as a valid piece of scientific evidence demonstrating that the highest temperatures of the last 1000 years have only occurred as a result of human addition of CO_2 to the atmosphere. It was so severely compromised that nearly all anthropogenic climate scientists recognized that they had better shift their discussion of observational evidence from the hockey stick to something else, such as the Vostok ice-core data.

The Climategate controversy has resulted in book-long accusations and counter-assertions between the contrarians and the anthropogenic activists and is far too time-consuming to detail here[161].

But frankly, the controversy over the hockey stick has no real significance anyway. It simply takes too short-range a view. The modified hockey stick attempted to show that the current warm spell was the warmest in the last 1000 years. But all one needed to do was look back at an earlier period (using a valid and consistent proxy) and one could not only find a warmer time, but could see that such temperature rises as we currently are experiencing are not unusual at all. (See graphic 10.3).

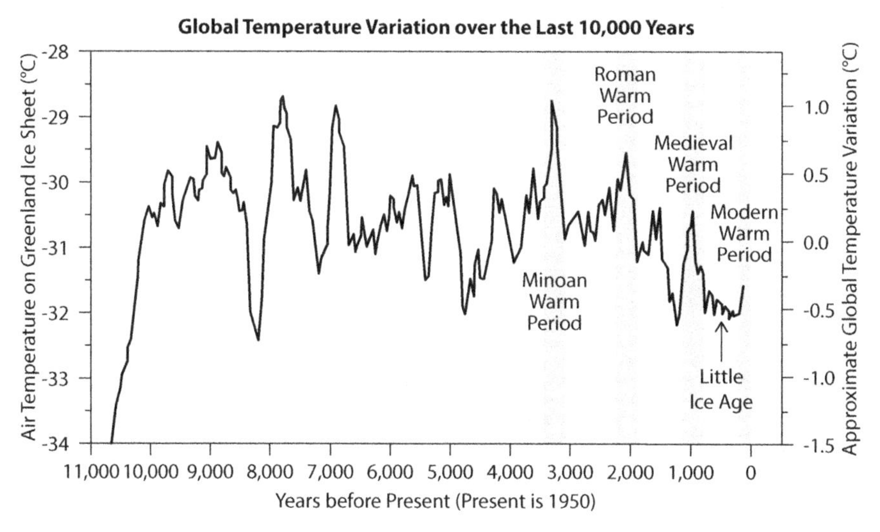

Graphic 10.3 **Global Temperature Variation, Last 10,000 Years,** R.B. Alley[162]. [Time →].

The multi-century temperature "spikes" such as the Minoan, Roman, and Medieval, warm periods shown in graphic 10.3 are obviously not caused by humankind's greenhouse gases (the significant emission of which began in the late 1700s). So what good does it do to attempt to attribute the Modern Warm Period (the last one to the right) to humankind's production of carbon dioxide and then be unable to explain the other warm periods?

The approach used by the inner core was to deny the existence of the medieval warm period and the Little Ice Age – which proved to be a bad idea since there was so much ancillary evidence of those two periods. And they ignominiously ignored the earlier warm periods (the Minoan and Roman and earlier "spikes" on graphic 10.3 – apparently because this kind of evidence did not fit their premise that humans and greenhouse gases were the principal causes of global warming.

Chapter 11

The "Evidence" Becomes Invalidated, But the Movement Keeps Trundling Along

Nothing is more damaging to a new truth than an old error.

— Johan Wolfgang von Goethe, 1749-1832

THE HOCKEY STICK SEEMED important to a tight group of anthropogenists whose outlook was apparently limited to an extremely narrow global view as well as a very short-term time frame. So, in addition to the theoretical evidence, the hockey stick observational evidence had now been severely discredited and was effectively out of the ballgame. The inner core did not publicly acknowledge that anything was amiss as they quietly retreated from this hockey stick mess and moved-on. Instead they began to emphasize the Vostok ice-core observational evidence (graphic 3.2) – with the now-coupled atmospheric/oceanic computer climate models as back-up.

It should be noted that at this point in the early twenty-first century, most of the inner core was apparently not aware of the emerging invalidation of the theoretical evidence that backed-up their premise of greenhouse gas culpability and thus they continued to believe that "the observational evidence should fit, if only we tried harder".

Thus, many of the just-outside-the-inner-core climate scientists continued to believe in their pre-conceived (but fallacious) foregone conclusion that global temperature was regulated by atmospheric greenhouse gas content – rather than welcoming the fresh, compelling contrary evidence that was emerging from the study of the sun. They could and should have been beginning to look elsewhere for the true causes of the global warming we were witnessing. But this is apparently what happens when scientists (who are urged on by environmental activists), become enraptured with a "cause" before enough data is available to draw a proper conclusion.

The last piece of popular talking-point observational evidence – the Vostok ice-core data also soon fell from grace, for even though the data were correct, the interpretation of those data was wrong. When the Vostok ice-core data (graphic 3.2) first appeared, most of the climate-science community immediately assumed that the increasing atmospheric CO_2 was causing the earth's temperature to rise, for in this case there appeared to be a strong cause and effect relationship. How else could there be so many matching synchronous up-and-down cycles?

Initially, there was only one major problem: no one could figure what was causing the atmospheric CO_2 to rise and fall – which seemingly then caused global surface temperature to rise and fall. The best paper on the subject, after discussing many alternatives, termed it "one of the great mysteries of global warming."

The Vostok ice-coring was a three-nation project. The Russians drilled, the French analyzed the ice-core data and reported it, while the Americans supplied funding and occasional field work. Finally,

Barkov, the lead Russian scientist of the ice-coring operation conducted a detailed cause and effect evaluation of the results. He closely scrutinized a stretched-out graph of the data showing ice-core temperature and also atmospheric CO_2 versus time (graphic 11.1). With this expanded view, he saw that the rising temperature curve occurred *before* the rising atmospheric CO_2 curve by hundreds of years.

Historical Image: Vostok Ice Core Data Analysis

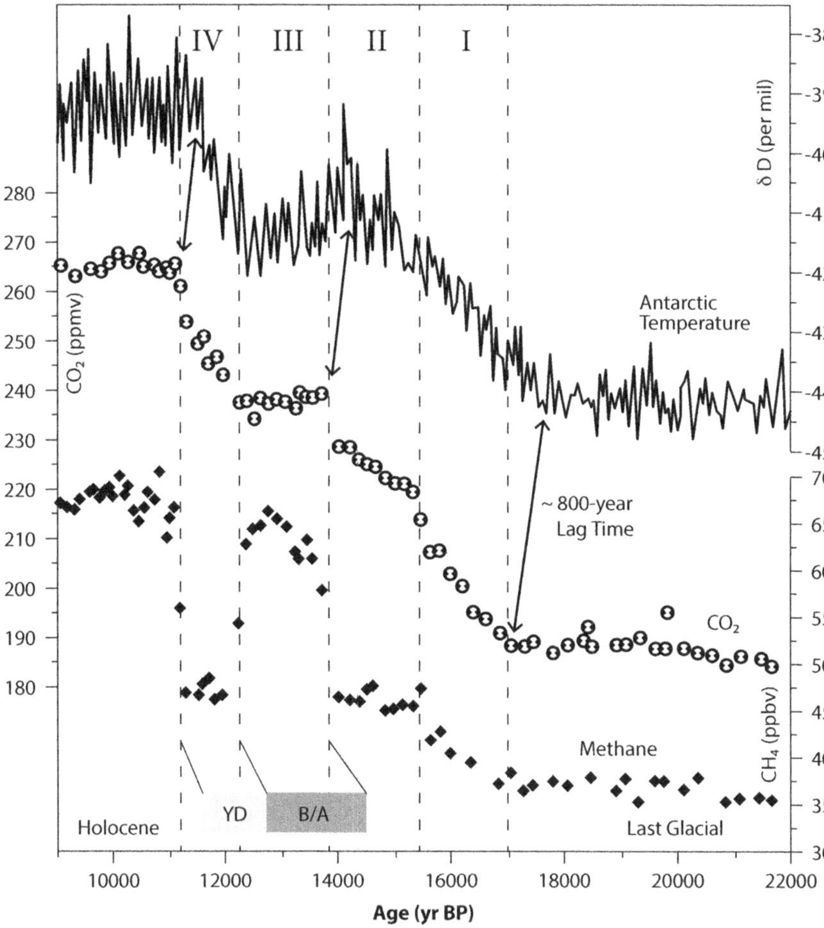

Graphic 11.1. Historical Image. **Vostok Ice Core Data.**, E. Monin, et al[163]. This telling image demonstrates that rising atmospheric CO_2 is a result of rising global surface temperature rather than a cause. [Time ←].

This lag between rising CO_2 content and rising temperature meant that, yes, there was indeed a cause and effect relationship, but that rising global surface temperature – which was caused by some other forcer (perhaps the sun) – was causing the ocean surface to slowly warm. The ocean, due to slow ocean current circulations and upwelling, in-turn, caused the deeper dissolved CO_2 to off-gas hundreds of years later – not as an amplifier, but as a non-effecting "indicator" (or "tracer") which happened to show up incidentally in the trapped gas bubbles in the deep ice-cores hundreds of years later.

Thus, the curve of increasing CO_2 started several hundred years *after* the curve of increasing global surface temperature. Apparently unrecognized by the inner core, some aspect of the sun (the most likely cause for a significant temperature change over this relatively protracted time frame) was making global surface temperature rise. The increasing atmospheric CO_2 (that off-gassed from the ocean as it warmed) was the following, but non-effecting result (or "tracer"), several hundred years later.

This finding – that the rising CO_2 was a result of the warming ocean rather than the cause of it – had to have been distressing to the anthropogenic inner core. It was antithetical to their most profound beliefs. They put a number of "rescue teams" on the problem. Three of these teams were led by some of the inner core's most prestigious scientists. They apparently were charged with the task of trying to "spin" things around so that the evidence could somehow be interpreted to confirm their anthropogenic version of what was causing the global warming (that is, increasing atmospheric greenhouse gases emitted by human activity). If they lost the ice-core as evidence like they lost the theoretical basis as well as the hockey stick evidence, then they would be down to just the computer models. And they knew these climate models were more assumption-filled, speculative, and uncertain than the observational evidence (and very much dependent on the preconception bias of the modelers and how they framed their conceptual package of inputs to the computer model).

Thus, with the establishment of these rescue teams that were headed by some of their leading scientists, this was an all-out effort to see if the facts could somehow be interpreted to fit the foregone conclusion. Otherwise the premise that humans were the principal cause of global warming would be in dire jeopardy. And losing the notion that humans were responsible for global warming had by now become a completely unacceptable option for the founders and propagators of this now strongly committed anthropogenic global warming "movement" with its hundreds of millions of followers. It was inconceivable to them that their human-caused atmospheric greenhouse gas version of events was incorrect.

During the following decade of trying, each of the rescue teams attempted to demonstrate that carbon dioxide could somehow be interpreted to be responsible for this long-range warming of earth – not the other way around. Their results are discussed in appendix C (for those interested in the details). Here we will simply say it should be obvious that the long-term changes in global temperature are the result of the earth's gradual changes in distance from the sun as a result of its changing elliptical path around the sun (as defined by the Milankovitch cycles). This meant that changing greenhouse gases were the *result* of changing solar heat received by the earth – and not the cause – just as the theoretical evidence (the physics) and the hockey stick both indicated. This will become quite clear later on in this book. The determined effort to pin changing global temperature on the changing content of atmospheric carbon dioxide in the atmosphere demonstrates how focused and fixated on atmospheric causes the anthropogenic global warming inner core scientists had become. They apparently felt compelled to make the evidence "fit" their hypothesis instead of stepping back and looking at the overall situation to find a hypothesis that "fit" the evidence.

A problem with these "quiet repositionings" is that the remainder of the scientific community as well as the journalistic, environmental, and general public communities never were alerted to either the move away from the unacceptable piece of evidence or the reason for the move. No one had admitted that anything was wrong with the original position apparently because the inner core didn't believe that there was anything wrong with their original position. They had simply not yet found a way to demonstrate that their original position was sound – as their very tenuous speculative evidence supporting that position steadily deteriorated.

Regrettably for the anthropogenic inner core, virtually all of the prime observational evidence that CO_2 was the principal cause of rising global surface temperature had now either become invalidated or was being seriously questioned.

With the deterioration of both the theoretical (physics) evidence and all the popular observational evidence (both the hockey stick and the Vostok ice-core data), the inner core had no choice but to try to pin their case on their last recourse, the computer climate models.

The Computer Climate Models

Our computer climate models can't account for the rising global surface temperatures we have seen over the last half century or so without factoring in human contribution to atmospheric CO_2[164] became the final defense of the inner core of climate scientists for their position that humankind's carbon dioxide was the principal cause of global warming.

Candidate parameters for a highly complex computer model generally fall into one of four categories:

- The parameter has been conceptualized (identified, isolated, and characterized); mathematically defined; and a computer with enough capacity to compute it is available.
- The parameter has been conceptualized and mathematically defined, but no computer with enough capacity to compute it is available.
- The parameter has been conceptualized but it can't be mathematically defined.
- The parameter has not yet been conceptualized

Only parameters in the first category will be in the computer model.

Billions of dollars were provided by the U.S. government so that climate computer modelers could obtain supercomputers with vastly increased computing capacity to allow parameters in the second category above to now fall into the first category and thus be used in the computer calculations.

With the successive acquisition of more than 20 exceedingly expensive very powerful and sophisticated supercomputers by the climate research community, the climate computer model community was able to make their model parameters successively "finer." They reduced the size of grid cells (some modelers call them "boxes") – making them sharper through each generation of climate assessment reports[165]. Attaining this increasing sharpness led these computer climate modelers to believe that they were getting more and more accurate results when they were really getting results of the same accuracy (and certainty), but they were more and more precisely depicted. The accuracy (and certainty) of the information was still erroneous because some of their input data was fallacious – and mathematically manipulating erroneous data does not make it more accurate, it simply makes the modeler think so.

The latest Department of Energy model under development and scheduled to be operational in the early 2020s, is considered to be "ultra-sharp" with grid cells only 25 kilometers (16 miles) wide – still too coarse to study individual clouds or atmospheric currents but a step finer than the existing grid cells[166]. But all this fine tuning is not a step forward except in the mathematical sense. To get a more certain answer (one that is closer to the actual truth) one must improve the accuracy (certainty) of the overall model (including the input package) as well as improve the precision. And you can't do that when such factors as the amount of global warming that takes place as a result of a doubling of atmospheric carbon dioxide are based on fallacious suppositions. This increasing computer power is making the field more precise but is not improving its accuracy (certainty). Unfortunately, too many people don't understand the difference, and think their answers are getting more certain when they are in fact, not. (See the clarification between precision and accuracy below).

Precision versus Accuracy

For those who do not fully understand the difference between the terms, precision and accuracy, picture the following scenario:

You are at a rifle range. There are ten shooting lanes, and each has a target on white paper with 10 concentric circles that are spaced one inch apart. You are in lane five. You shoot eight rounds and all hit the bulls-eye (the innermost concentric circle). Your friend is in lane three. His eight rounds spray over the entire face of the target. Your precision is excellent, but your friend's precision is not so good. But the range-master then announces over the bullhorn that, "Number five" (that's you) "has been shooting into the target in lane six". So, while your precision is still excellent your accuracy is way off. Your friend's precision was much worse than yours but his accuracy was much better than yours. In other words, your friend would have grazed his victim's cheek instead of hitting his forehead, while you, instead of hitting your victim, would have hit his lawyer.

Before improved precision can help, you must have good accuracy. But as we saw earlier, in the anthropogenic global warming premise, the inputs are "off" by a factor of about five. Accordingly, the accuracy (and certainty) are "way-off" before you have even gotten started. No amount of improved precision is going to help your accuracy or your certainty. Thus, all this effort to improve your precision is just a tremendous waste of resources – until you remove the fallacious inputs and substitute correct input data.

This is one of the fundamental problems with the whole approach to the science of attempting to verify that humankind causes global warming by using computer models with erroneous input data packages. The enraptured have made a vast industry of peripheral or secondary "climate research" with thousands of scientists busily consumed by problems that use fallacious information as their starting point. Accordingly, their conclusions are invalid and no matter how much they manipulate that incorrect input data, the results will remain invalid until they correct their input data.

The Computer Results Bite the Dust

A different example of the futility of relying on the computer models is shown in the Midlands chart. (See graphic 11.2 repeated here from chapter 1). The notion that the computer climate models can't explain the recent rise in global surface temperature without using human emission of CO_2 into the atmosphere centers around the postulation that we are currently seeing a more extensive and more rapid change in temperature than would be the case if it were caused by natural factors. This is shown in the right-side short "uptick" trend line. On the other hand, the left-side short "uptick" trend line was not apparent to the anthropogenic global warming inner core of climate experts because they did not consider data from that far back in the past.

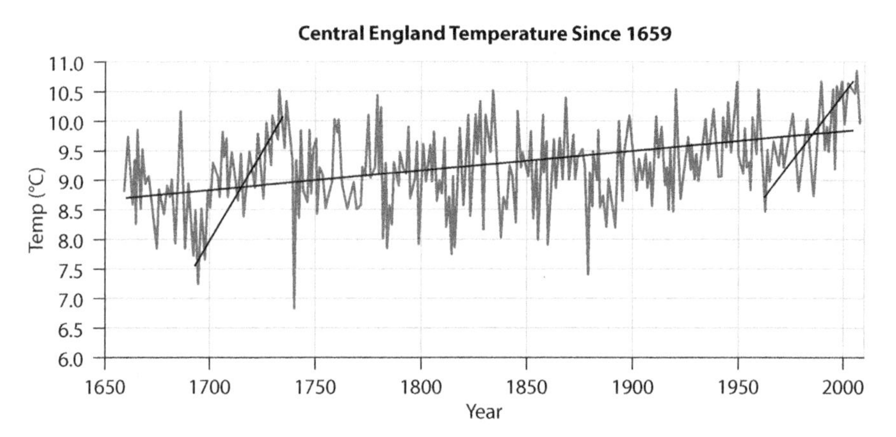

Graphic 11.2. **Temperature – The Midlands of England** adapted from G. Manley et al[167]. (Repeat from chapter 1)., The longest reliable single-point instrument (thermometer) temperature record in the world. [Time →].

This left-side "uptick" must be the result of natural factors since it occurred in the 1600s and early 1700s before the onset of the Industrial Revolution (the late 1700s) and thus before the onset of the widespread use of fossil fuels (anthropogenic causes). It is more rapid and of greater extent than the right-side "uptick" trend line. If the only way the computer models can explain the rapid rise and extent of the right-side trend line is to include humankind's atmospheric CO_2 in the calculation, then whatever parameters caused the left-side trend line, which shows greater rapidity and greater extent than the right-side one, are not in the computer model – and therefore such natural activity could be the cause of the right-side trend line (without any help from humankind's CO_2.) So the computer models have just proved themselves to be incomplete or just plain wrong. Clearly natural factors are causing the rapid change in global surface temperature on the left and could be, as well, on the right. (These natural factors will be discussed shortly).

The computer climate models, despite their massive capabilities, were and still are, simply no match for the far more massive (and becoming better understood every day) complexities of the earth's multi-faceted natural climate variability – many features of which have not even been conceptualized much less mathematically defined by the anthropogenic climate computer modelers who, by-and-large, have not yet come around to looking at possible causes other than atmospheric greenhouse gases.

Chapter 12

Wind and Solar Energy

Things are not always what they seem...

— Plato (Phaedrus), 370 BC

WIND AND SOLAR ARE renewable sources of energy and have been accepted by the anthropogenic global warming environmentalist community as the "answer" to our energy plight. The thesis was that humankind had to stop using our principal energy base of fossil fuels just as soon as possible and that "nonpolluting" wind and solar, along with other renewables (possibly supplemented by nuclear power), would be the replacement.

So political actions were taken to spur the development of the renewable replacements for fossil fuels by incentivizing and subsidizing them. The growth of wind and solar had become a high priority to many environmental activists in many countries. Sympathetic governments (led by the United States) spent billions of dollars on subsidizing the implementation of wind and solar.

As of 2018, a little more than 3% of the world's electric energy came from wind turbines and about 1% from solar panels (both photovoltaic and concentrated-thermal combined). And that use is accelerating as governmental policies are incentivizing the use of these kinds of energy because of the perception they do not contribute "harmful" carbon dioxide to the earth's atmosphere, are otherwise "pollution-free", and would make excellent low-cost replacements for our fossil fuel base as the world decides to decarbonize.

Unfortunately, wind and solar both have variability and intermittency problems which are hardly noticed when they are used just as minor fringe supplements to a continuous reliable base supply (as they are now in most areas), but become a massive problem when we attempt to use them as a replacement for the fossil fuel base. *This is because the electric distribution grids have a built-in "reserve margin" of from about 12.5 percent to 22 percent excess capacity to cover unexpected outages and demand peaks. And fringe use of wind and solar can impinge on this reserve margin relatively innocuously if use is minor.* However, if we envision wind and solar being anything more than a fringe supplement – such as becoming the new base, then that's a whole new ballgame. To plan an energy future based on wind generators and solar panels would see a dramatic cost increase from what we estimate today based on their use as minor supplements to an existing base. This is because of the mismatch between time of generation and time of use and the need to provide energy storage or stand-by (or some other means to bridge the gap) in order to provide a reliable supply.

Anthropogenic activists are impressed with the economies-of-scale and the learning curve savings that are being reported in the production of wind generators and solar panels. They seem to think that this theoretical savings trend on these manufactured generating products is going to apply to the entire generating system until wind and solar become truly competitive with the fossil base – and at that point the full-scale conversion to renewables will become economically viable without government subsidy. But unfortunately, that will not be the case at all – for a number of reasons:

First, the projected learning curve theoretical savings are not being realized over the entire system because "unanticipated" expenses have arisen when those generators are incorporated into a functioning electric system [168]. Most wind and solar "farms" are located where the wind and sun are and also where there is room for them, not necessarily where the customers are. Thus, transmission lines are often much longer and expensive than for power systems currently in use.

Secondly, the land area required for these farms is much greater than for a traditional concentrated power plant, giving wind and solar a much lower "power density ratio"[169] which will become more and more expensive as wind and solar energy penetration of the base-load increases and land area available for such use becomes farther and farther away from the energy consuming area. One of the reasons professional energy analysts don't take renewables very seriously is because they know that in the long run – if the policy makers of the world tried to convert a considerable portion of our energy from fossil to wind and solar – an unacceptably large portion of the land area of the entire world would be covered with wind generators and solar panels. Opposition to locating these wind and solar sources of energy close to people – even in relatively low population areas – is already beginning and can be expected to grow[170].

Third, and most importantly, the mismatch of the time of generation to the time of use requires some provision to bridge the gap. The obvious first choice is energy storage either in the form of storage-battery banks or pumped-hydraulic (where water is pumped into a higher reservoir during periods of low demand, and allowed to flow back through water turbine driven generators during periods of high demand). Unfortunately, both batteries and pumped-hydraulic are costly – but are not required when use is minor and the excess capacity of the reserve margin is being used instead, as is now the case.

Interconnection to adjacent grids is an often-mentioned alternative to energy storage. But we are talking about decarbonizing – so the adjacent grid will also be powered by wind and solar. And when looked at carefully, more wind and solar, even in a slightly different geographical area, is simply not reliable enough to act as a backup for existing wind and solar sources. In a nutshell, we are talking about an unreliable source backing up another unreliable source, and that simply does not work. (See appendix E for a more complete explanation).

Another hoped-for remedy – combining a gas-fired boiler with a wind or solar generator is an often-discussed potential remedy. And it works when penetration of the baseload is small. But, since the intention is to consequentially diminish the amount of carbon dioxide in the atmosphere as penetration of the baseload increases, this option doesn't work either, because it emits considerable CO_2 into the atmosphere from the backup fossil fuel, natural gas – and again we are trying to decarbonize.

As stated above, initial small-scale use of wind and solar can impinge on the grid's reserve margin without being noticed and thus the need for (and cost of) energy storage or other provision to bridge the gap between the time of generation and the time of use is not apparent because the reserve margin is absorbing the need for extra power when there isn't enough. But as penetration of baseload deepens, it becomes necessary to replace the reserve margin. Now the true cost of these intermittent wind and solar generators should begin to show-up. But maybe not, because of an interesting concept proposed by anthropogenic activists, called grid parity.

Grid parity is a subtle concept designed to hide the cause of the mismatch of wind and solar time-of-generation with time-of-use and the resulting need for energy storage provision to accommodate that mismatch. This is accomplished by looking at all providers of power on an equal basis without incorporating the factor that current baseload providers (coal, gas, nuclear, and hydroelectric) are essentially available when needed while wind and solar are frequently not (and grid management has no control over when they will be). Then, when it becomes necessary to add energy storage provisions or additional generating capacity (as wind and solar penetration of baseload increases) under the grid-parity concept, the cost for such provisions is applied to all the providers, not just the providers (wind and solar) that caused the need for it – thus disguising that wind and solar are the cause and not identifying them as the reason for the price increase. A very shrewd concept.

Even many utility executives have succumbed to the activist pressure of the movement and are advocating grid parity by joining the "anthropogenic global warming go-along-to-get-along movement" and accepting and endorsing what is really nothing more than an attempt to make wind and solar look more economical than they really are when all aspects of cost are considered.

As wind and solar penetrate further and further into the baseload the cost will go up and up and up. And that is the scary part of the whole concept of increasing wind and solar to a point beyond it being a fringe supplement.

There is strong evidence that it will be prohibitively expensive to make the transition from fossil fuels to wind and solar renewables on a scale needed to consequentially stem the flow of carbon dioxide into the atmosphere while still maintaining a dependable supply of electricity.

Sometimes the need for dependability is underappreciated. But electric power outages of today involve far more than having to cook dinner over an outdoor charcoal broiler with a flashlight like we did some decades ago. During the "great outage" of August 2003, a cascading failure of the U.S. eastern electric lattice caused some 50 million people in eight northeastern states, to successively lose all power. The cities were particularly hard hit as tens of thousands of people were stranded in elevators, subway cars, airports, and commuter trains for many hours. Most of the fast food facilities that supported these transportation stations had to close down (for lack of power) just as the need dramatically increased. Traffic lights were extinguished and city traffic jams proliferated, shattering nerves and tempers. Most city apartment dwellings have no charcoal barbecue grill and no alternate lighting or air conditioning. In some, windows don't even open. In these modern times in well-developed countries such as the United States, we are highly dependent on a *reliable* source of electric power.

In Great Britain, which has a fairly advanced and geographically diversified wind-generating capacity, there were some 152 occasions in a recent two-year period when the total wind generated over the entire wind system dipped to below 2 percent of nameplate capacity. The weather became so quiet in June of 2018, that Britain went nine consecutive days without being able to generate any wind power at all.[171] When penetration of baseload is deep and demand is high, and this happens, even if we have a very expensive interconnection, can we really depend on grid B to have sufficient excess capacity to support grid A, as well? The concept of solving the unreliability of wind and solar through interconnection of power grids across wider geographic areas has not been fully thought out and is not a practical solution to the intermittent availability problem. (See appendix E).

Another option is to simply dismiss the importance of matching the *time of generation* with *time of use.* And many theoretical reports and plans match the amount of power generated to the amount of power needed (which can honestly be done for a fossil, nuclear, or hydroelectric generating plant because they can be turned on when needed), but doesn't work for a wind or solar farm because they quite possibly will be generating the most when it is needed the least and sometimes not generating at all when it is desperately needed. The cheapest option of course is to just let the supply fade away when the wind dies. But then you do not have an effective continuous power supplying system. For solar, of course, you get a complete die-off every evening just when you need it the most.

In South Australia, wind has become a consequential component of electric power and as of 2017 it supplied nearly 40 percent of demand. South Australians already have the highest electricity costs in the world, paying about three times what electric power costs in the U.S. with its fossil fuel base[172]. Yet they are constantly plagued with fade-outs and interruptions (some of which are caused by weather events such as high winds and lightning strikes). To try to approach a solution to this unreliability problem, South Australia, in 2017, contracted with Tesla and Elon Musk to provide a $50 million battery bank capable of supplying some 30,000 people (out of 1.7 million on that grid) with a hopefully reliable "fade-free" supply. Clearly this additional $50 million investment will not come close to completely solving the problem. And as penetration of the baseload by wind increases from the current 40 percent (if it increases), the need for more and more battery banks will continue to escalate the already high cost of electricity in South Australia. Additional battery banks just to address their current problem could approach $2 billion (or conceivably much more depending on the downtime engineers assume) in additional expenditure without adding any additional capability other than possibly partially solving the current fade-away problem.

With the above as an example, one can see how the price for wind and solar power rapidly increases with baseload penetration. When thought through, these energy-storage costs are really very high and must be added to the already high renewable generation system costs. And they grow and grow as the percentage of generation moves away from the reliable steady base of fossil fuel, nuclear, or hydroelectric generation to the variable and intermittent replacement – "renewable" wind and solar.

Countries such as Germany and particularly Britain have struggled to meet idealistic but unrealistic reduced-carbon commitments they have made to their own people as well as the rest of the world. Proponents of the British Climate Change Act stated it would add the equivalent of "the cost of a scoop of ice cream" to monthly energy bills. Instead it has squeezed the very poorest in the UK to the point of total disbelief and loss of confidence in promises made by government since their electric bills have increased dramatically[173].

As stated above, it is possible for wind and solar to penetrate the baseload up to about 15 to 20 percent without really noticing the need for price increases – until it is realized that the reserve margin has been usurped and must be restored. Then all the projected savings disappear as prices soar.

Various other suggestions for reducing the need for energy storage to match time of generation to time of use for wind and solar have been made. Some of the ideas, for instance, suggest an apparently well-thought-out approach that utilizes demand management; dispatchable generators; forecasting and planning; (and other ideas), and could be used in many instances to help reduce the need for energy storage[174].

But some suggestions to fulfill the need for energy storage, which might include anything from heat stored in rocks underground; pumped hydroelectric; modified hydroelectric dams; batteries; flywheels; compressed air; heated salts; and/or hydrogen disassociation[175], all come with an apparently unrecognized high price tag. Any attempt to store electric energy is going to be expensive. And the more wind and solar penetrate the energy baseload use, the higher the energy storage cost becomes. Vague ideas are easy to come by, but practical capital projects are often much more expensive in reality than when conceptually (and incompletely) dreamed-up.

For example, how much battery reserve is required for the project? Eight hours? Twenty-four hours? The storage battery requirement (and cost) goes up and up as the predicted down-time-span increases. Who would have prognosticated nine days? And maybe even that isn't enough. Can you imagine the cost of a series of storage-battery-banks able to cover an outage for a significant portion of a major grid for a time-frame of nine days? Engineers for Fukushima Daichi nuclear station in Japan used eight hours – and that was just enough power to run the main coolant loop pumps to prevent a core meltdown. Unfortunately, it was far from being enough and the worst nuclear disaster in history resulted with perhaps 16,500 lost lives.

As California's penetration of baseload by wind and solar rose to 23 percent in 2017[176], the need for energy storage became evident. They had used up all the available reserve margin and the grid could no longer respond to peaks that the margin was supposed to cover. In addition, the energy surplus that was available when both wind and solar peaked while demand was low had no useful outlet. It was time to restore the reserve margin. So, the State of California is fostering a $2 billion proposal to use Lake Mead (formed by the Colorado river at Hoover Dam in Nevada and Arizona) as a pumped hydroelectric reservoir by adding downstream pumps and pipelines to pump the released water back up into Lake Mead so that it can be used again to "even the load"[177]. As California closes down fossil and nuclear plants in their zeal to convert to all "renewable" energy they are beginning to have to pay the piper. Will the people of that state recognize that this $2 billion is an added cost due solely to increasing wind and solar penetration of base-load? Or will it simply be buried as a part of grid parity and its cost allocated to all the other generators of electricity?

There are so many unknowns in the wind and solar route that it is very difficult to properly plan and engineer the system as the level of penetration of baseload increases. Do we really want a national electric lattice where, when the wind and solar penetration is only up to 40 percent of baseload, the cost per kilowatt hour has already become some three (or so) times what the present fossil fuel base costs? Yet also have fade-outs and outages that are much more frequent and of unpredictable duration? These are the features of the present South Australian system, (and, to a certain extent, the British and German systems as well) and similar results can be expected for a North American system. And imagine how much greater the cost per kilowatt hour will be when the penetration of baseload by wind and solar rises to, say, 60 percent or, heaven forbid, 80 percent.

Recently, unexpected erosion and corrosion problems have been discovered in wind turbines. It seems the leading-edge of the wind-blade is subject to "micro-particulate" erosion that was not accounted for in the original design. Older wind turbine blades are experiencing unacceptable leading-edge erosion which takes them out of service far earlier than their originally expected design life. Also, wind farms on off-shore but shallow ocean locations are experiencing increased corrosion that their designers did not anticipate. Both of these developments indicate a shorter life for these forms of electric energy generation than originally calculated, with attendant higher than expected costs.

Increased use of wind and solar renewables would also have significant long-term consequences that we tend to sweep under the rug, namely slowing the earth's rotation and changing the world's climate by changing wind patterns and intensity. These problems increase from the almost undetectable (when use is minor as it is now) to more and more detrimental as penetration increases – with the ultimate consequences growing over time.

The High Shelton wind farm near Buffalo, New York interferes with TV and radio reception. It also interferes with the local weather forecasting radar (and even disturbs the local weather), making it impossible to have accurate weather forecasts in that area. These are not the kinds of problems expected by the public from the movement toward renewables. And when you multiply these problems by thousands of times (if we were to move toward a significant transition to wind and solar, they demonstrate that considerable chaos would result.

Many in the electric utility advanced-planning field who are not enraptured by the anthropogenic "movement", feel that we should skip over or at least minimize the wind and solar route. If we remove the artificial incentives (such as subsidies and tax abatements) from these uneconomical and potentially environmentally damaging methods of supplying electric power, then they can wait for the day (as fossil fuels slowly deplete) when they may be found acceptable as the only available alternative, despite their many drawbacks.

You can see that we, as a people, must prioritize the development of controlled thermonuclear fusion as the replacement for fossil fuels which we currently expect to deplete sometime in the 400 to 1000-year range (See Epilogue). The ITER project, led by the Europeans is "nearly there" in developing a successful prototypical controlled fusion reactor. But the U.S., as a minor participant, is not leading the charge as it should be. Fortunately, others are, and we'd better begin to notice, or we are going to be left in the tailwind of this difficult but extremely important advancing technology.

Our national research priorities are dramatically misplaced. We should be planning to make an orderly transition to the almost-certain successor to fossil fuels – controlled thermonuclear fusion – and then to its natural successor, the more advanced process of elemental annihilation, as the reliable new sources of energy into the future. Although they sound ominous, both of these sources are essentially very much safer than nuclear fission (our present nuclear power plants) because neither fusion nor annihilation requires a hot radioactive core that must be continuously cooled and is subject to meltdown and subsequent explosion.

In the United States, wind and solar have become a financial windfall to select (mostly foreign) investors bolstered by the largesse offered by a generous U.S. policy of subsidy for alternates to fossil fuels brought on by the fallacious conclusion that carbon dioxide emissions must be eliminated at any cost or we are all doomed. These renewable energy subsidies actually waste taxpayer money on projects that are of no real value to anyone other than those investors.

Problems multiply with wind and solar renewable energy as their penetration of baseload increases. Many policy-makers world-wide and most of the public just don't seem to realize this.

There is no reason to continue financially subsidizing wind and solar. When subsidization stops, the growth of both wind and solar will also stop (until fossil fuels become considerably more expensive than they are now). The carbon dioxide produced by the use of fossil fuels is beneficial to our atmosphere. So fossil fuels can continue to be used until fusion is ready to step up to the plate and become the energy source for electricity of the future. There is no need for rushing fossil fuels out of the picture. They will eventually deplete over the next 400 to 1000 years or so (or perhaps even sooner).

This is not to say there will not be "changes of use". For example, electric automobiles as well as electric domestic heating should be encouraged in high-population areas that have air pollution problems. The electricity to power those autos and heaters can still be produced by fossil fuel powered generators that are located in areas where the exhaust is more readily controlled by precipitation, filtration, or dissipation.

You might say that wind and solar were yesterday's knee-jerk response to the last decade's naive perception of the energy dilemma, not today's answer to a more enlightened assessment of the future energy paradigm.

Chapter 13

The "Movement" versus the "Science"

Two roads diverged in a wood and I – I took the one less traveled... and that has made all the difference.

— Robert Frost, 1916

THE SCIENTIFIC UNDERSTANDING OF complex magnetic waves within the sun and their influence on earth's sustained cloudiness (and resulting reflectivity of incoming solar radiant heat), which we will discuss shortly, was becoming understood by a very few solar-oriented scientists. Most of the inner core of atmospheric climate scientists and their increasing number of followers in not only the climate-science community but even more importantly, in the downstream peripheral scientific disciplines, were not at all tuned in to this improved understanding of the sun and its effects on global climate. They had what they believed to be "the answer": humankind's greenhouse gases were causing global warming and its effects were well enough understood to be quantified (even if that quantification was based on suppositions that were now being scientifically contradicted). This meant that peripheral scientists, statisticians, economists, bloggers, alarmists, journalists, politicians, worriers and speculators had something to work with. The fact that it was becoming apparent to some scientists that it all was fallacious was simply ignored by the enraptured. For them it seemed so easy to understand:

- Carbon dioxide is a greenhouse gas that retains heat.
- Humankind is pouring huge quantities of carbon dioxide into the atmosphere.
- The earth is getting warmer.
- Therefore, carbon dioxide is causing the earth to warm.

Now they thought they could predict the future by making a few assumptions about the rate of continued human use of fossil fuels. This told them projected future atmospheric CO_2 content and thus future global temperature could be easily calculated. From there, multiple, diverse, adverse environmental and human effects (such as a rapidly rising sea-level) could be prognosticated. Unfortunately, because of a simple scientific misinterpretation two decades earlier, the climate scientists had arrived at incorrect figures for how much influence rising greenhouse gases had on global temperature. And since these invalid figures were what the downstream scientists were using, they were calculating faulty results, drawing fallacious conclusions, and creating a widespread fictitious world climate change illusion.

And what an in-place funding bonanza! The billions of dollars in research money available each year from some thirteen U.S. government agencies (whose budgets are largely established as an inflated continuation of last year's budget) had already begun making an orderly shift from attempting to demonstrate that the earth was indeed warming (by various proxy means) to the many harmful effects this now "established" human-caused warming (or climate change) would have on the environment and on humankind. So, this was a natural continuation of that already-begun transition in the changing direction of climate research.

As we progressed into the twenty-first century, the consequences – the grim warnings – appeared to get more and more ominous. The latest *National Climate Assessment* that was issued in November of 2018 showed us the climax of all this specious speculation[178]: a ten percent decline in GDP; devastating effects on human health; an environmental disaster with increased wildfires, crop failures; disrupted trade; increased drought in already dry areas; increased rain in already wet areas; rampant heat waves with $141 billion in heat related death costs… and on it went. There was only one problem. It was all based on a fallacious supposition to begin with – that human caused greenhouse gases were responsible for global warming and that future warming could be predicted by estimating humankind's future use of fossil fuels. It was all a total climate change illusion with no basis.

An important unfortunate point, however, was that all that "scientific" effort resulting in the accumulation of some 1656 pages of findings in the *National Climate Assessment* made most people inside the anthropogenic-enraptured climate community actually believe that these ominous outcomes were becoming more certain simply because of the increasing amount of accumulated so-called "scientific evidence" (all of which was sheer speculation). More and more reports, more and more people involved, more and more "believers", and more and more pages of (fallacious) documentation – all based on a false starting-point assumption – did not add up to anything but a series of imaginative misleading climate change illusions (and a tremendous waste of resources).

So, while a few solar scientists were homing in on the real causes of the global warming the earth was experiencing (the multiple solar factors that we will discuss shortly), the people within the anthropogenic global warming "movement", led by the atmospheric climate scientists, were convincing themselves (and the public) that they were further verifying their "humankind and its greenhouse gases are causing climate change" conclusion.

Since it was all based on a false supposition, of what value will this research be? It now appears that a sustained mid-term time frame global cooling process (which apparently is really controlled by recently recognized magnetic forces within the sun and their relationship with the earth) will start in earnest in less than a decade.

The 2007 IPCC Fourth Assessment Report Detection & Attribution section[179] declared that they were now:

- 100 percent sure that increasing atmospheric CO_2 was caused by humankind.
- 100 percent sure that the globe was warming.
- 90 percent sure that humankind's CO_2 caused most of the warming of the last half century.

Knowing what you know now, you might think this jump from 66 percent to 90 percent conclusion stretches the upper limits of incredulity. But you must remember the fundamental thinking of the IPCC summarizers. Not only does this group not understand the CO_2/water vapor "overlap misinterpretation" that makes their version of the anthropogenic warming physics invalid, but virtually all of the information showing the unraveling of the observational evidence and the computer model evidence for the anthropogenic premise simply does not exist for them. This is because the IPCC, which controls the climate change information dissemination platform, refuses to report any part of anything that conflicts with their anthropogenic foundational charter. Thus, even many of their own members are unaware of these recent "contrary outcome" developments.

Likewise, the emerging evidence for the natural causes (solar proximity to the earth and solar magnetic participation in earth's changing cloud cover and reflection of incoming solar heat – which we will discuss shortly) also does not exist in their sphere of understanding. Recognizing the huge scientific implications of this massive information deficit, perhaps the best way to understand this belief of an advancement from a 66 percent sure to a 90 percent sure conclusion that humans and greenhouse gases are responsible for climate change, is to think like an anthropogenic-global-warming climate computer model scientist for a moment:

Remember that today's computer-driven climate-change science involves beginning with a conceptual package of facts and "assumed truths" as data inputs – an ensemble of which is then mathematically treated in a computer so that some values can be successively changed, and then finally an assessment is concluded. During this computer math treatment, the modeling scientist may come to believe that the statistical level-of-confidence of the computer math operation reflects the level-of-confidence (the certainty) of the overall answer – when it really reflects the level-of-confidence of just the computer mathematical operation. The level-of-certainty of the overall answer is dependent on the certainty (the accuracy) of the various components of the initial inputted conceptual data package as well as the subsequent mathematics: *You don't add certainty to a faulty as-*

sumption by mathematically manipulating it.

So when you hear or read "We are 90 percent sure" by the IPCC, what that really means is: "We are 90 percent sure *if our input of assumed and judgmental data is entirely correct"*. And since we know in this case that at least one of the fundamental inputs is completely incorrect (the 50/50 split in heat absorption influence of water vapor and carbon dioxide in the infrared absorption band overlapped by both), then we know that the conclusion is also fallacious, even if the computer math is pretty good.[180]

Yet eager adherents to the anthropogenic "cause" seemed to accept that somehow in the intervening five or seven years between the Third and Fourth IPCC Assessment Reports, as the inputted data has been found to be more and more doubtful, they have gone from 66 percent sure to 90 percent sure that humankind is responsible for global warming when such a conclusion is completely absurd. Perhaps you could say the precision improved from 66 percent to 90 percent while the accuracy (and certainty) went from 35 percent to 10 percent (this is strictly an example).

The stated basis for the move from about 66 percent sure in the IPCC Third Assessment Report to about 90 percent sure in the Fourth Report was "expert judgment". In this case that means the judgmental assessment of the summarizer. And the statistical level-of-confidence of the computer's calculation was, of course, an important factor. Thus the "expert judgement" was really a product of the summarizer's preconception and the prevailing conventional wisdom. It was evident that the IPCC Detection & Attribution summarizers were marching to the drumbeat of the "public opinion guidance" arm of the IPCC.

The IPCC Fifth Assessment Report of 2013/14 stated that: *We are now 95 percent sure (extremely likely) that human emission of green-*

house gases into the atmosphere is the principal cause of global warming[181]*. This was up from 90 percent (very likely) in the Fourth Report and 66 percent (likely) in the Third Report. While to the enraptured, the case was becoming more and more confirmed and more and more ominous with each new report, to those who understood the correct science, the scenario was becoming more and more absurd and more and more ridiculous with each report.*

As in the case of the IPCC Fourth Assessment Report, the strength of the Fifth Report with its 95 percent sure conclusion had clearly come from the confidence level of the computer math operation, not the certainty of the overall bundle of information. The unfortunate consequence is that there are virtually millions of people all around the world who simply do not understand that this whole scientific perception is based on a supposition that stems from a scientific misinterpretation made by the inner core that failed to account for the heat-absorbing role of water vapor over that of carbon dioxide in a small overlapped portion of the infrared spectrum back in the 1990s. Since that time, virtually thousands of peripheral scientists who started their investigation with an incorrect premise, have been using a supposition that is scientifically false, having been based on that misinterpretation. And they have been using it as a scientifically established fact and as the conceptual basis (unquestioned starting point) for their research apparently without challenging it or making any effort to affirm it. And furthermore, why should they, when "everyone" seems to believe it, and there is plentiful grant funding available for those who do believe? It is an example of another "informational cascade" whereby an unproven opinion of a few, starts to "snowball" into a widely-accepted and widely-used "law of the Medes and Persians" despite being based on fundamentally flawed information.

The reason that some climate change books with a contrary theme seem so confrontational (even antagonistic) in their approach is that these authors understand the absurdity of the anthropogenic position without understanding how the anthropogenic community got to that position. Thus, they believe purveyors of that notion are charlatans and fraudsters. However, those purveyors did not arrive at their false conclusion by a series of malicious and penurious actions with the intention to consciously deceive the public – they were on a legitimate quest to understand the world around them. Unfortunately, as they were getting millions of people to believe their story, the evidence backing up that notion was deteriorating. It was then that they apparently felt they could not "let-down" their growing mass of followers, and thus could not bring themselves to go back and check their input data. When you are winning, keep running.

As stated earlier, apparently many of the anthropogenic inner core and their closely associated colleagues are so shielded from reality and influenced by their own public information (public opinion guidance) section of the IPCC, that they are simply unaware of the importance of much of the new information that has been emerging in this field. The inner core and the IPCC summarizers, perhaps fearing what they might discover, unfortunately do not look beyond the realm of atmospheric causes and do not see any reason to check their original assumptions to see if those assumptions are still valid. Their environmentalist supporters often restate "The science does not change", despite the massive amount of evidence that the science is frequently being fleshed out and reinterpreted. Yes, the science does change (or at least the human understanding of the science changes). And in the field of climate change, the science has materially changed during the last two decades – a period that carries us from the time of the *proclamation of anthropogenic cause* at the turn of the twenty-first century to the end of the second decade of the twenty-first century.

Climate scientists seem to be influenced by and are simply following the thinking of their closely associated environmental activists with their intransigent but obsolete beliefs. Instead, climate scientists should be leading the activists with the very important emerging science. As it is now, with their simple reshuffling of outdated and fallacious information, they are not helping the field of science or anyone else.

Chapter 14
Solar versus Greenhouse Gas Forcing

There is no belief, however foolish, that will not gather its faithful adherents who will defend it to the death.

— Isaac Asimov

Over the years, "climate-forcing"[182] graphs and charts were published by the inner core that show considerable global temperature forcing by atmospheric gases and little forcing by solar heat radiation. These charts helped shape the inner core's thinking at the time of their *proclamation of anthropogenic cause* in January of 2001. Anyone referring to one of these radiative-forcing charts would assume that greenhouse gases have a significantly greater influence on global warming than the sun does, when actually, just the reverse is true. The following three charts pretty much demonstrate this long-held erroneous belief by the inner core of climate scientists:

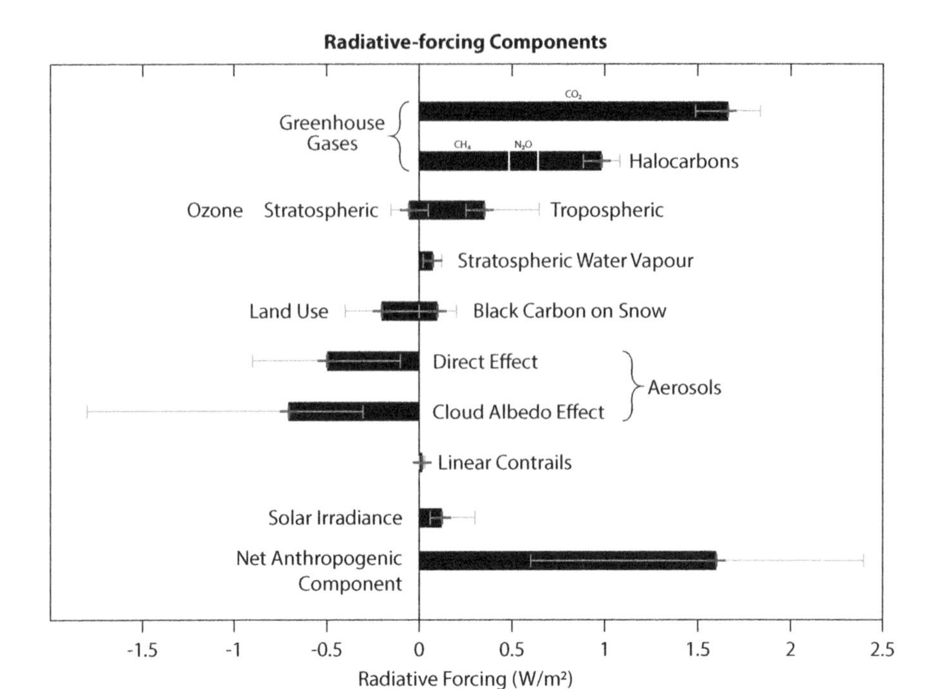

Graphic 14.1 **Radiative Forcing.** Adapted from Leland McInnes[183] (Repeat from chapter 1). This chart is typical of the many different ones produced over the years by the inner core of atmospheric climate scientists – virtually all of which showed a tiny amount of solar forcing and a large amount of atmospheric greenhouse gas forcing.

This graphic (14.1), and its preceding and successor "updated" charts is probably the most disappointing series in the entire arsenal of information produced by the anthropogenic global warming scientific community. It has been endorsed by some of the most prestigious members of that group, yet it still is a good example of what is wrong with so much of climate science. Both the math and the chart rendition are seemingly excellent. The slight changes in new generations of these charts have swiftly passed peer review from the very best peer reviewers who apparently have become inured by its presence over the years. It is an "old friend". The problem is in the input data that the very first chartists in the series began with. The information in those input packages, although perhaps conforming to the incorrect information available at that time, is dramatically distorted based on the correct information available now, as you will see in appendix B. The essence of these early charts has been passed-down through the succession of reports and graphs on this topic starting in the mid-1990s. The same basic misinformation has been carried through despite the radical differences that have emerged. These graphs all show considerable forcing by greenhouse gases and much less (in fact almost none at all) by the sun. These widely-viewed preceding, derivative, and successor charts are probably the prime reason for the perpetuation within the scientific community of the deceptive notion that the changes in solar forcers do not have enough power to cause the changes in temperature that we are witnessing on earth.

We will now ignore the distractive extras shown on this particular chart and create our own chart showing just the solar versus primary greenhouse gas information in that chart that directly applies to this discussion: (Be sure to read appendix B for more detailed information.)

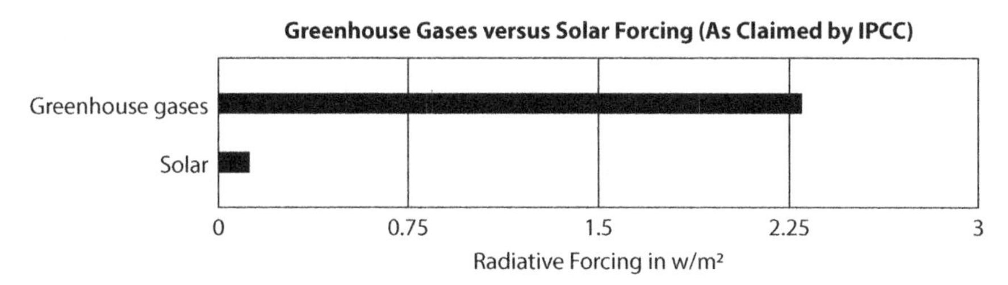

Graphic 14.2. **Principal Greenhouse Gases versus Solar Forcing (As Claimed by IPCC)** without distractive information. (From Appendix B)[184].

Note that this chart clearly depicts the major influence on global warming by the principal greenhouse gases and very little by the sun, which is a main tenet of the anthropogenist position.

Next, we'll update the errors in both bars to show the proper influence:

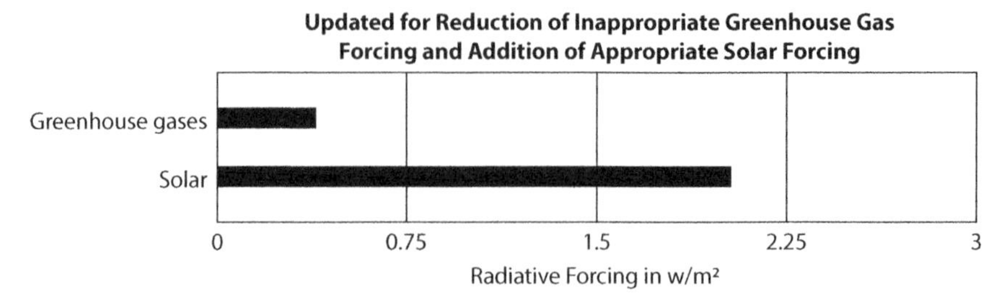

Graphic 14.3: **Graphic 14.2 Updated for Reduction of Inappropriate Greenhouse Gas Forcing and Addition of Appropriate Solar Forcing.** (From appendix B)[185].

You can see that the updated chart now shows almost the direct opposite of what the earlier chart showed. The scientifically valid corrections are detailed in appendix B and show the reasons why the greenhouse gas bar was far too long and the solar bar was much too short in the original series of charts. If there is any doubt in your mind about this very important point be sure to read appendix B. This set of two charts completely turns around the misconception that solar factors are not powerful enough to cause the global warming we are witnessing while greenhouse gas forcing is. After reading Appendix B, you will see that greenhouse gas forcing is not sufficient to cause the global warming we are witnessing while solar forcing is.[186]

At this point it should be becoming clear that human-caused, fossil fuel-related, greenhouse gas increases are not the principal cause of global warming and climate change. At first there was the suggestion that human responsibility might be a valid consideration until the physics clearly showed it wasn't. Next, the observational evidence quietly deteriorated. And finally, it became evident that the computer climate model evidence was telling us whatever the computer modelers wanted it to say – and thus simply reflected their anthropogenic bias.

Where has climate science taken us in the past three decades? Two decades ago, a select few climate "experts" could understand only one possible cause of the global warming we were witnessing – humankind's emission of greenhouse gases into the atmosphere. They candidly admitted that they "weren't sure" that was the cause of global warming, but since it was the *only cause* they could envision it was better to tell the public (and policy-makers) that it was the principal cause than to risk not having said anything and later have it turn out to be true. Sort of a lesson learned from the *great Galveston hurricane* of 1900 when the forecasters knew of the possibility of the storm but didn't say anything because they didn't want to unnecessarily alarm the public. Some 8000 people lost their lives as a result of the forecasters silence and inaction. This time the climate scientists thought the result would be far more ominous. So the time for a crucial decision for the climate science community came in the late 1990s. The inner core of climatologists decided that since the scale of the problem was potentially so vast, they'd better tell the public there was an established danger (when it really was only a possible danger) in order to induce them to stop using fossil fuels – just in case it might later prove to be true[187].

Now (in the early 2020s) we are hundreds of billions in research and subsidy dollars later with literally tens of thousands of scientists, journalists, environmentalists, and politicians operating from a starting point assumption that they are convinced is based on sound, rock-solid science – but is really based on what was a very scientifically questionable judgmental supposition. A supposition that has since been found to be invalid.

All the prognostications of doom dreamed up by a legion of anthropogenic disaster-thinkers who all started with a false supposition (that the globe will continue to warm until we stop using fossil fuels) will be shown to be false because they are based on a faulty premise. All the horror stories of continued warming are based on what we now know to be incorrect assumptions. Carbon dioxide is a clear, colorless, odorless, gas that is beneficial, in fact necessary for the existence of plant life. Its presence in the atmosphere is very near the lower limit for sustaining plant life. We should be putting as much carbon dioxide as is possible into our atmosphere. And best of all, even though it might have a trivial effect, its tiny presence has no influence of consequence on global temperature or climate.

Now we are left with the realization that the earth is warming, but what was thought to be the only possible explanation for this warming by atmospheric climate scientists – humankind's fossil fuels emitting greenhouse gases into the atmosphere – is not actually the cause of this warming.

So, if it isn't humankind, and fossil fuels, and greenhouse gases, what is it that's causing the earth to warm?

The work of the solar-oriented scientists has begun to come to the forefront. Telescopes of much greater optical power as well as those with various non-optical sensing capabilities have come into use. Multiple specialized satellites of a myriad of various types have begun to tell scientists a lot of new things about the sun and its multiple properties that have turned out to have formerly unsuspected temperature influence here on earth. In parallel with all this new detecting equipment, new supercomputers and their vastly expanded capabilities also have come into play. New methods such as wave analysis, synoptic mapping, and principal component analysis have opened up advanced methods of looking at the newly detected data. A whole new field of science is quietly emerging. This has led to discoveries about the sun's magnetic properties that have a significant effect on earth's magnetic shield – which influences earth's cloud cover and regulates the amount of the sun's heat that is reflected away from the earth by clouds (and thus strongly influences global surface temperature).

Previously unappreciated – and in some instances unsuspected – complex radiational as well as magnetic properties of the sun that have a profound effect on the earth's surface temperature are finally becoming recognized, identified, characterized, and contextualized.

Change is the process by which the future invades our lives.

— Alvin Toffler

Part II
The Real Causes of a Changing Global Temperature

Eureka!

— Archimedes, c 225 BC

In Part I of *The Climate Change Illusion* we showed that we can be reasonably sure that human-caused emission of greenhouse gases into the atmosphere is not the principal cause of a warming earth. But that leaves us with the unanswered question: "If it isn't the human use of fossil fuels, what is causing the earth to warm?" Answering that question is the subject of Part II.

Chapter 15
The Magnetic Sun

O, Sun…

— William Shakespeare

RECENTLY, MUCH HAS BEEN learned about the internal workings of the sun and its various "outputs" that influence global temperature. The array of solar "forcers" that has recently become recognized as important cannot be dismissed by atmospheric climate scientists simply because they previously had a "*low level of scientific understanding*". Solar scientists are far from having all the answers, but they are beginning to appreciate that the sun and its changing multiple forcers as well as the relationship of those forcers with earth's complimentary "moderators" (deflectors and reflectors) are the principal regulators of earth's changing temperature and climate during mid-range time frames (multiple-hundreds of years).

Solar-Magnetic and Earth-Cloud Relationships

Clouds are formed from droplets of condensed water vapor. Water vapor needs a particulate nucleus to condense onto when atmospheric conditions (temperature and pressure as well as ample water vapor) are ripe for the formation of water droplets. These droplets then form into a cloud. Aside from earth-originated particulates (forest fires, volcanos, desert storms, spores, smoke stacks, exhaust pipes, and so on), a large portion of these droplet nuclei come into the atmosphere from the sun in the form of the solar wind of free atomic nuclei with some of the electrons stripped and of loose protons and electrons which collide with atmospheric gas matter like nitrogen and oxygen atoms to create particulate "showers". Also, galactic cosmic rays enter the earth's atmosphere from all directions and act similarly. These solar and galactic particulates (both the solar wind and cosmic rays) are "charged-particles" and are subject to attraction, deflection, or repulsion by magnetic fields. Thus, any changes in the magnetic shield surrounding the earth, influences the arrival of such particles into the earth's atmosphere by changing their direction. *The sun's magnetic field is so strong that it heavily influences the net magnetic field surrounding the earth (the earth's so-called magnetic shield, which is made up of a combination of the earth's magnetic field and the sun's magnetic field). Accordingly, solar caused slow changes in this magnetic shield consequentially influences earth's sustained cloudiness.*

The extent of earth's cloudiness influences how much solar heat is reflected away from the earth. This makes clouds an important regulator of the earth's heat-balance. A sustained change in earth's overall cloudiness will have a consequential influence on global surface temperature and climate. And changes in earth's cloud cover are strongly influenced by the changing solar magnetic field (which then changes the earth's magnetic shield).

Thus, sustained variations in the sun's magnetic field strongly influences global temperature and climate. (This point cannot be emphasized enough). Yet this very important global warming dynamic was not considered by the inner core of climate scientists when they formulated their *proclamation of anthropogenic cause* in January 2001 that attributed global warming to humankind's emission of greenhouse gases into the atmosphere as the principal cause of global warming. You will recall that one of their prime reasons for so concluding was that climate scientists were not aware of any other alternative. But, as has just been pointed out, there is indeed another very viable alternative. *In addition, as also has been pointed out, the climate scientists' explanation of presumed cause (changing greenhouse gases), has since been found to have, at best, only a trivial influence on global temperature and climate, and cannot be responsible for the kinds of global temperature changes we have seen and are currently witnessing.*

Earth's albedo (reflectivity to incoming solar heat) is primarily controlled by earth's sustained extent (or degree) of cloudiness.

The Major Recent Solar Revelation

For some years, solar scientists have been studying the solar dipole magnetic field reversal of about 9 to 14 years, averaging about 11 years (22 years for the complete cycle) which, as well, is of significant but slow varying intensity. In addition, the concept of a solar dynamo and/or double-dynamo is generally accepted[188]. *Many solar scientists have been trying to correlate this eleven-year magnetic reversal dynamic to changes in solar surface heat output – and there is some correlation. But by-and-large, up until very recently they seem to have been missing the most significant factor – which is the changing solar magnetic effect on earth's magnetic shield and its effect on earth's cloudiness (and thus on the amount of solar heat reflected away from the earth – and thus, earth's surface temperature).*

Recently a certain mid-term rhythm and regularity (multiple-hundreds of years) to this varying solar magnetic intensity has been identified and characterized (in addition to the basic 11/22-year cycle). There now also appears to be a *dual-magnetic wave differential phase progression cycle (we'll call it the solar dual-magnetic wave cycle)* within the sun[189]. Although some models show the dual-magnetic waves generated from interior longitudinal differential (armature-like) rotation within the sun[190], others are now showing the differential rotation being sectional. The sun rotates on its axis like most galactic bodies. Being a viscous-plasma blob, the sun experiences differential-rotation wherein massive sections of the sun rotate at slightly different rates around its central axis[191]. Starting at the sun's equator and heading toward its north pole, sections of the sun travel at different rotational speeds. Near the equator it takes 25 days to travel one revolution. Then various sections (or discs) take 26 days, 28 days, 31 days and finally near the north pole 35 days. Heading south from the equator there is a similar differential rotation except the rate at the south pole is slightly different from that at the north pole. Thus the rate of rotation at the equator is faster than at the poles and the rate at the two poles is slightly different from each other. Combined with the interior flow of "rivers or currents of plasma", this makes for a very complex magnetic dynamo. For illustrative purposes this can be simplified by lumping these multiple adjacent viscous-plasma discs into three viscous-plasma sections that will make two magnetic waves that are slightly out of phase with each other. Graphic 15.1 is a depiction of this simplified magnetic scenario.

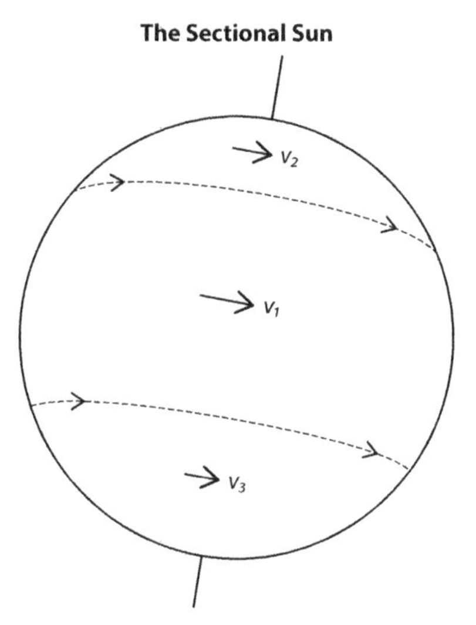

Graphic 15.1: **The Three-Section Sun**. Depicting three viscous-plasma sections rotating at slightly different velocities (*v1, v2, and v3*)[192].

For purposes of this discussion we are lumping together sections of the sun that are really a series of continuous amorphous masses. But depicting them as three discreet entities simplifies the discussion: The sun's equatorial section rotates slightly faster (*v1*) than its two polar sections which also rotate very slightly out-of-phase (*v2*) and (*v3*) with each other[193]. The sun also has massive crossing slow-flowing sub-surface currents. All of this sets up a dual-magnetic wave differential phase progression effect. Thus, there are dual magnetic waves that, due to the sun's differential rotation between the northern and southern hemispheres, are slightly out of phase with each other. When these dual-magnetic waves slowly come into phase with each other, the total magnetic strength is reinforced and strong. As they go slightly out of phase the combined magnetic force is progressively less strong – and when they arrive at opposite phases, they tend to reach a minimum.

Thus, the sun's *combined dual-magnetic wave phase progression cycle* – the sum of the fields of the magnetic waves (which we are calling the solar dual-magnetic wave cycle) – rises and falls on a slowly pulsating multi-hundred-year cycle (or beat-frequency). It takes multiple solar bi-polar *magnetic field reversal* cycles for the dual-magnetic waves to phase-orient from completely in-phase to completely out-of-phase. The current minimum (cancellation-phase) to maximum (addition-phase) solar dual-magnetic wave cycle is presently estimated to take about 34 solar magnetic reversals (17 full cycles) or about 375 years for the half-cycle (and 750 years for the full cycle). It appears this cycle can vary considerably over time for different solar dual-magnetic wave cycles – as the partitioning of the sectional masses of the sun changes during the differential rotation process, as edge clusters (or masses) of viscous-plasma transfer or change-over from one solar section to the adjacent one.

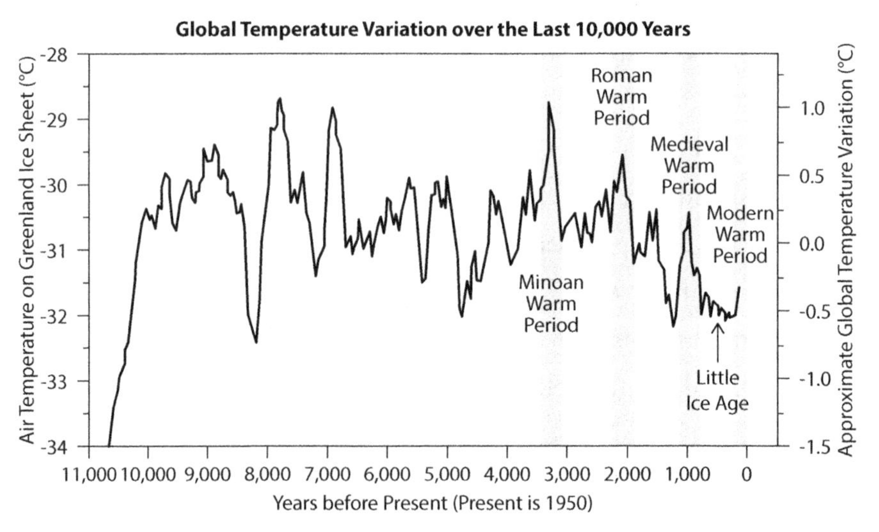

Graphic 15.2. **Global Temperature Variation last 10,000 years,** R.B. Alley[194]. (Repeat from chapter 10). [Time →].

Remember, these combined solar magnetic field changes strongly influence the earth's magnetic shield which, in turn, influences the earth's sustained cloudiness and its reflectivity to incoming solar heat radiation – and thus have a consequential effect on changing global temperature and climate even during times when the solar irradiance or solar surface heat output is relatively steady. It is postulated that this solar magnetic *dual-magnetic wave cycle* is responsible for most of the temperature peaks and valleys that occur on earth in the multiple-hundreds to a thousand years or so range that are evidenced in the ice core, marine sediment core, and other time/temperature data. Graphic 15.2 shows multiple temperature peaks or "prongs" over the last 10,000 years of earth's history that are best explained by this *solar dual-magnetic wave cycle.* In addition, shorter-term "slips" and "cluster mass" transfers of plasma between the basic differentially-rotating sections of the sun may be the cause of shorter-term temperature "upticks" or "downswings" in the multiple-decade (half-century) range.

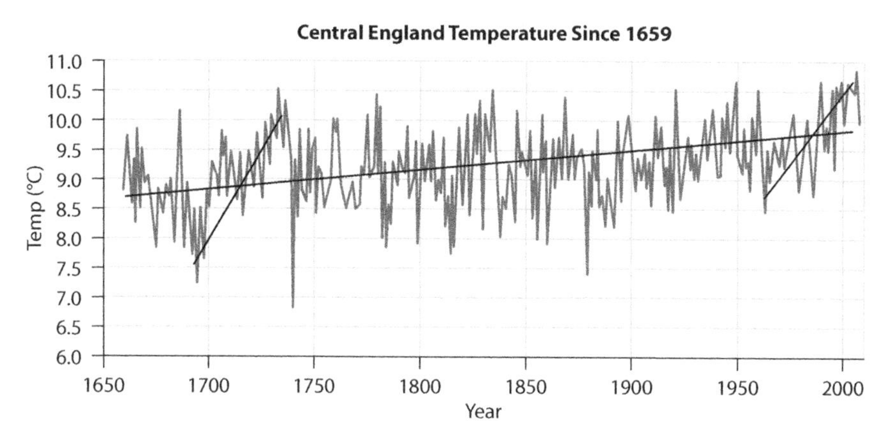

Graphic 15.3 **The Midlands of England. Temperature since 1659,** G. Manley et al[195]. (Repeat from chapter 1). [Time →]

On a more spread-out scale, the global temperature rise we are going through currently, as we approach a peak (or solar dual-magnetic wave in-phase maximum) is shown in graphic 15.3. This mid-term time frame increase in global temperature (smoothed by the central linear trend line) shows a definite gradual temperature increase ever since the Little Ice Age in the late 1600s.

Irregularities in the heat radiation output from the sun (caused by solar storms, etc.) may cause many of the minor irregularities seen in graphic 15.3. But remember this is a monthly average so there is also some seasonal influence. The more abrupt global temperature shifts (as shown by the two uptick trend lines in graphic 15.3) may be caused by not-yet-well-characterized "cluster transfers" between the sun's various viscous-plasma regions rotating at slightly different rates. As stated above, it is postulated that during this differential rotation, sometimes large clusters or "slabs" of viscous-plasma mass break away from one section and join the adjacent section. This might produce a more rapid temporary magnetic shift that, in turn, causes a temporary (up to five to seven decades or so) change in earth's average sustained cloud cover with a resulting change in reflectivity of incoming solar heat radiation and a resulting five to seven decade more rapid temporary change in global temperature. Larger cluster-mass transfers (or "breakaways") would result in longer time frame solar section differential rotation changes which, in turn, would result in longer changes in earth's average sustained cloud cover – with resulting longer global cycles of temperature change. And this may explain past longer mid-term time frame temperature excursions (of up to a thousand years or so) as noted in historical temperature data such as graphic 15.2 and the finer detail in graphic 15.4.

The evidence has become quite strong that there is a *"mid-term" (multiple-hundred to a thousand years or so) changing solar magnetic field cycle that affects the earth's sustained cloud cover which, in turn, is a major regulator of earth's global surface temperature.*

More Than One Cause

The anthropogenic climate "experts" (the inner core) had settled on a "one-cause-fits-all" explanation of events wherein changing atmospheric greenhouse gases (primarily carbon dioxide, methane, and nitrous oxides) was the cause of all global temperature changes past and present and for short-term, mid-term, and long-term time frames. This was because they could envision no other possible cause for fluctuations in global temperature of the types we were seeing over those different time frames. So they had to make the only potential cause they could see fit the global temperature changes detected for all time fames as shown in various proxies such as ice-cores, marine sediment cores, and cave speleothem, as well as the more current instrument (thermometer) and short-term proxy records.

But now, it was becoming evident that there was more than one cause of changes in global temperature. And these different causes were applicable for different time frames. For starters, the long-term (100,000-year cycle) depicted in graphic 15.4 was clearly *orbital* and defined by the Milankovitch cycles. The changing distance (proximity) between sun and earth that was experienced during these long-term orbital cycles was enough to cause the global temperature changes that were noted in the Vostok ice core data that approximated 12°C. There was no longer any need to try to attribute these temperature changes to changing atmospheric gases. The distance (proximity) change between the earth and sun experienced during this lengthy cycle easily explained the global temperature change. The effect being much like that of moving your hand away from a hot stove – the temperature of the stove isn't changing, but the amount of heat your hand receives from the stove does change because of the change in proximity.

In addition, the magnetic and radiational interrelationships between earth and sun were beginning to be understood as having influence on mid-term time frame global temperature changes. The field of study was widening despite the inner core's refusal to expand their parameters and look at anything other than atmospheric gas causes. They had ruled out the sun early on by looking exclusively at changing heat radiating from the sun's surface. This turned out to be only one of the multiple potential causes of global temperature change from the sun (and a minor one). They ruled changing solar irradiance "not powerful enough" to cause the temperature changes they were seeing and incorrectly concluded the only other possible option was changing greenhouse gases on earth. Thus, they fostered (and accepted) the *proclamation of anthropogenic cause* of January 2001, in which humans and greenhouse gases were considered the principal cause of global warming and climate change.

What the atmosphere-oriented climate scientists were unaware of was that the solar influence on earth's surface temperature actually consisted of multiple components, not just the one considered (but underestimated) at the time of the *proclamation of anthropogenic cause* of January 2001. These multiple components of climate forcing by the sun include:

- Changing solar surface radiant heat output. (Most noticed in the shorter time-frames).

- The changing proximity of the sun to earth. (Only detectable in the exceedingly long-term [paleo or orbital] time-frames).

- The sun's changing magnetic field. Most importantly, the changing solar *combined dual-magnetic wave phase differential progression (or solar dual-magnetic wave) cycle* – which is responsible for mid-term time frame (multiple-hundred to a thousand or so years) temperature changes.

- The exchange of edge-cluster slabs of viscous plasma mass between the solar differentially rotating sections and the resulting more rapid (sometimes five to seven-decade) consequential magnetic changes that contribute to global temperature changes in that time frame.

- The changing rate of solar particulates (the solar wind) – which influences earth's cloud cover and solar heat reflection in ways still being evaluated. This is being extensively studied by NASA. See appendix F.

Many theoretical solar scientists think we are very near the current peak of the solar combined dual-magnetic wave phase progression (solar dual-magnetic wave) cycle that has been going on since the Little Ice Age some 370 years ago – and therefore we soon should begin to see "global cooling". Thus, the central trend line of graphic 15.3 should soon begin to tilt slightly downward instead of continuing the current upward trend. Likewise, the final prong on graphic 15.2 (which first must be extended to show the increase in temperature from 1950 to 2020) should then also begin to turn downward since it shows the same information as graphic 15.3 on a far more compressed time scale.

Thus, virtually all of the climate research now being conducted by thousands of scientists in other disciplines (such as biologists) on the effects of continuing global warming (who are using the fallacious assumption that the cause of global warming is increasing atmospheric greenhouse gases) will be rendered immaterial when they realize that they are using the wrong parameters to predict future global temperature.

> *Some 120 scientific papers in 2017 alone linked climate change to variations in solar internal and surface activity and its determinants in the reflection of radiation process (clouds, solar wind, etc.). It is becoming increasingly established that such changing solar activity causes mid-term (multiple-hundreds to a thousand or so years) temperature changes on earth*[196] *-- and that the recent phase has been one of warming (and is about to turn to cooling).*

Yet none of this is acknowledged by the "establishment" atmospherically-oriented proponents of anthropogenic global warming and climate change (the inner core of climate scientists) who are entirely focused on changing atmospheric greenhouse gases and how the earth will continue to warm until humans curtail their use of fossil fuels. It recently has been clearly demonstrated how these changing greenhouse gases have only a trivial effect on global temperature (about five times lower than that estimated by atmospheric climate scientists[197]).

Coincidentally, during the last 150 to 200 years or so of this solar magnetically induced global warming period which has been going on for about 370 years (the current solar dual-magnetic wave half-cycle), humankind's emission of greenhouse gases was also gradually increasing due to our increasing use of fossil fuels. Obviously unaware of any of this *solar-magnetic, cloudiness, reflectivity* process being the dominant cause of the global warming that earth was currently experiencing, atmospheric climate scientists who were under pressure to declare the cause of the then-current warming, drew the fallacious conclusion that the only possible cause of that warming had to be the result of *changing greenhouse gases.*

The full explanation of the solar-magnetic process has only come to light very recently, and by the time it did, the whole world already seemed to be marching to the tune of the incorrect *anthropogenic-greenhouse gas* hypothesis.

Scientists have come a long way in studying the sun and its various magnetic and radiative processes (as well as in recognizing the effects of the sun's proximity to the earth) since atmospheric scientists decided they didn't know enough about the sun to include its influence in a changing global temperature or climate back at the turn of the twenty-first century.

Changes in solar heat irradiance caused by proximity (changes in distance between the earth and sun), are the prime cause of long-term (sometimes called "paleo" or "orbital") global temperature and climate changes as shown in graphic 15.4:

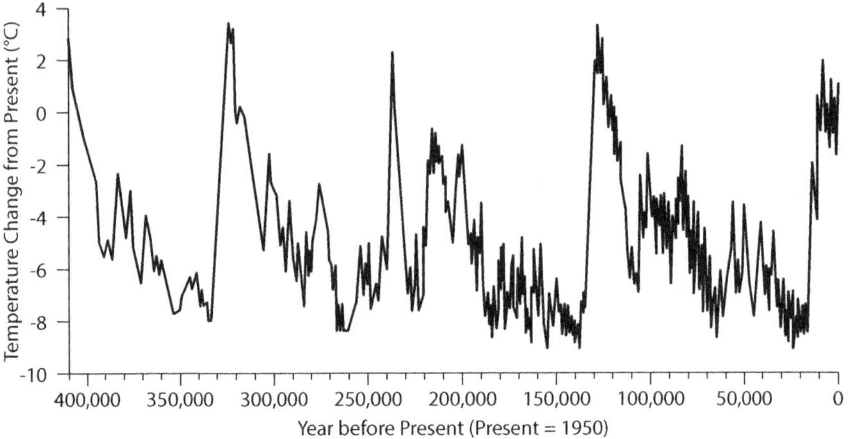

Graphic 15.4 **Global Temperature Anomaly Over 420,000 years,** from Vostok ice-core, J.R. Petit, et al [198]. (Repeat from Chapter1). [Time →].

Graphic 15.4 shows four iterations of the 100,000-year Milankovitch cycle. Note that it takes considerably longer for the temperature to decrease to the ultimate glacial minimum platform (or valley) than it takes to rise back up to the maximum genial platform (or peak). Apparently this is because the alignment of the other planets during the downswing part of the cycle is such that their combined gravitational influence pulls the earth away from the sun for a longer time; while during the upswing part of the cycle – when the earth moves more rapidly back up toward the genial maximum temperature platform – the reconfigured combined planetary forces are gravitationally pulling the earth more rapidly toward the sun.

The interaction of the many pre-existing momentum, inertial, gravitational, Coriolis, and centrifugal forces involved that regulate the earth's complex motion during this sophisticated cycle are ripe for more detailed scientific characterization, quantification, and integration.

There is still much to learn about both the solar changing magnetic field and the proximity of the sun to the earth and their influence on global temperature.

We now know that changes within the sun and in the earth-sun relationship are far and away the most important aspects of a changing global temperature and climate.

Changes in atmospheric trace greenhouse gases such as carbon dioxide, methane, and nitrous oxide have, at best, only a trivial influence on global surface temperature. Even if we used-up all the world's reserves of fossil fuels and emitted all the resulting trace greenhouse gases into the atmosphere it would have no consequential influence on global temperature or climate. (For more details on the magnetic and radiational relationship between the earth and sun see appendix F.)

Chapter 16
The Expanded Understanding

Then on the shore of the wide world, I stand alone,
and think...

— John Keats, 1820[199]

IN ADDITION TO THE solar effects, there are other external influences on the earth's temperature. The earth encounters changing quantities of space dust (micro-meteors) as it moves through space, and the earth's atmospheric absorption of these particles affects cloud formation. A "dust-disk" of space debris and fine particles in the earth's orbital path has been postulated as have random clouds of space dust, waiting for the earth to pass through as we move through space. It is estimated that sometimes up to 40,000 tons of space dust are encountered and absorbed by the earth's atmosphere in a single day. This dust influences earth's cloudiness and thus reflectivity of solar heat radiation. This source of cloud-making-nuclei is variable and largely uncharacterized and thus causes what today are considered unpredicted or random global temperature changes of varying duration and intensity.

And, of course, the particulate matter that originates on earth from forest fires, volcanos, dust storms, pollen, spores, sea salt as well as man-made particulates (smoke from various industrial and domestic processes) is variable and often causes unpredicted changes in global temperature. Neither of these two (space dust and earth-originated particulates) are charged-particulates and are not subject to deflection by magnetic fields – and thus are not subject to the rhythm of the changing solar dual-magnetic field cycle (although they could be subject to an as-yet uncharacterized "rhythm" of orbital space dust clouds).

Galactic cosmic rays coming from all parts of outer space that are entering the earth's atmosphere apparently also play a role in cloud formation. As charged particles, they are subject to the rhythmic deflection of the cyclic solar-caused part of the earth's changing magnetic shield (the combination of the earth and sun magnetic fields). Thus, they influence earth's changing sustained cloudiness. There is some debate among cosmic scientists about just how much cosmic rays versus the ionic solar wind influences cloud formation. But both are charged particles and thus both are influenced by the changes in the changing magnetic shield that surrounds the earth.

Although some of these other "factors-of-change" in the global climate are random or are not yet rhythmically characterized, at least two have been found to have regularity (in addition to the Milankovitch cycles):

- The 11/22-year solar magnetic field reversal cycle (which has some predictable connections to solar heat radiation) and is the current subject of intensive study by many solar scientists.
- The much slower in-phase/out-of-phase solar dual-magnetic

wave cycle of multiple hundreds of years to a thousand or so years (depending on the partitioning of the sun's differentially rotating viscous-plasma sections) with the current half-cycle projected to be about 375 years.

The solar dual-magnetic wave cycle is apparently the baseline of the relatively slow and gradual increase in global temperature we've noticed for the past 370 years and are currently witnessing as shown in graphic 15.3. While this shows as a gradual increase on shorter-term graphs (such as graphic 15.3), it shows-up as a "spike" or "prong" in mid-term graphs that have a compressed time scale (such as graphic 15.2). The sun-earth relationship matrix chart (graphic 16.1) may help clarify this relatively complex solar situation:

THE SUN-EARTH TEMPERATURE RELATIONSHIP MATRIX

A. *The output of the sun as it relates to the earth's changing temperature:*

1. Solar changing radiant heat output: ultra-violet; visual; particularly the near-infrared.
2. Solar changing ionic output: protons; electrons; stripped elemental nuclei (collectively called the solar wind). These are charged particles and are subject to deflection by the earth's magnetic shield. They create particle showers in the earth's atmosphere that become the nuclei of water droplets and thus help form clouds. The extent of these clouds then regulates global temperature by reflecting solar heat away from the earth.
3. Solar changing magnetic field: The combination of: (a) the magnetic field reversals (11/22± year); (b) the solar dual-magnetic wave cycle (multi hundreds of years or so); and (c) the rotational mass "cluster transfers" between solar sections (five to seven decades or so).

↑ ↑ ↑

The degree of influence of all the above is affected by the changing proximity of the earth and sun, which is very slow but is extremely influential over very long periods of time.

↓ ↓ ↓

B. *Earth's moderators (deflectors and reflectors) of the incoming output from the sun.*

1. Earth's changing magnetic shield – which is made-up of a combination of the changing solar magnetic field and the nearly stable earth magnetic field. This shield influences both the incoming solar wind, and incoming galactic cosmic rays.

2. Earth's rhythmically changing extent of cloudiness. This affects incoming solar heat radiation, which makes it a major factor in regulating global surface temperature.

Thus, the heat ultimately received at the earth's surface is determined by a combination of changing solar heat plus changing solar particulate radiation plus a changing solar magnetic field as modified by a changing pattern of earth reflectors and filters (such as the earth's magnetic shield and earth's cloud cover) surrounding the earth as well as the proximity of the two bodies. Some of the outputs as well as some of the moderators have "regularity" and already have been characterized while some need further study before they can be fully characterized, isolated, quantified, and time defined.

Graphic 16.1 **The Sun-Earth Relationship Matrix**[200]

As well, the atmospheric-oceanic "flywheel" influences on global weather and short-term climate have become better characterized even if these forces are sometimes difficult to explicitly predict. The short term (multi-annual to decadal) role of the Pacific El Nino/La Nina oceanic thermal-layer overturning is much better understood. Other oceanic thermal-layer overturnings and current-collisions (within the thermohaline undersea conveyor) are being used to explain weather and short-term climate changes. Oceanic multi-decadal overturnings (and "see-saws") are hypothesized.

All these relatively short-term, oceanic/atmospheric flywheel factors, while they have a regional effect and may even influence the apparent rate of change of global temperature, do not influence the overall magnitude of heat input to the earth, but do have a temporal effect and also do influence regional temperature change. In addition, they are indicators of weather or short-term climate variability. (For a brief comment on clouds see note [201] at the end of this book).

The earth's surface temperature changes we have recently witnessed can be explained by natural solar-influenced causes. There is no need to try to "force-fit" such changes to human-caused increases in greenhouse gases, which have only a trivial influence on global temperature.

So, while the evidence was deteriorating that anthropogenic *greenhouse gases* might be an important factor in global warming, the evidence that *solar-related factors* were the principal causes was increasing. In addition, at least some atmospheric climate scientists were finally beginning to accept that the very-long-range changing-proximity effect on global temperature caused by the Milankovitch cycles and the resulting change in proximity between the earth and sun during the very long elliptical cycles the earth took around the sun was the most important extremely long-range factor in determining the earth's surface temperature.

And the understanding of all this was emerging only *after* the decision by climate activists to "freeze the science" at the point of the crucial turn-of-the-century *proclamation of human-cause*, of January, 2001[202].

Observational Evaluation of Potential Causes of Global Warming

A Finnish researcher, Antero Ollila, recently looked at the effect of three potential causes of changing global surface temperature: (a) changes in atmospheric greenhouse gas content; (b) changes in solar irradiance; and (c) changes in reflectivity (cloudiness) over the last several hundred years to see if he could find a human fingerprint[203].

He found there was almost no correlation between atmospheric carbon dioxide or other greenhouse gas changes and global surface temperature changes, but that there was good correlation between global temperature changes and changes in the sun. Ollila also found there was cause and effect correlation between global surface temperature changes and cloudiness changes caused by space dust (mentioned above) and by changing ionic particle patterns both of which were influencing earth's cloud cover[204]. Thus, the influence of clouds, the important mechanisms of their formation, and the characterization of their role as a regulator of global temperature were all beginning to be fleshed out and more completely understood from this type of observational evidence.

Until very recently, few scientists were aware of the solar dual-magnetic wave cycle and its explanation of the mid-term time frame changes in global surface temperature that seemed so hard to explain. The addition of this feature contributed enormously to the overall understanding of global warming and the dominant role of the sun in this climate changing process.

Doubt

Some science historians have tried to sell the notion of anthropogenic global warming mainly on the argument that it is the scientific consensus[205]*. According to this narrative, the grater the apparent overwhelming consensus, the truer the premise becomes. Will we never learn? In this book, we don't challenge the notion that the overwhelming scientific consensus is that humans and their greenhouse gases are responsible for global warming. We reason that this scientific conglomeration is largely unaware of the misevaluations that were made in the formation and initial assertion of that premise, and that the same initial fundamental misinterpretations have been carried through, unnoticed, unexamined, and unchallenged from decade to decade for nearly a half-century of scientific discourse – both written and verbally conveyed. In this book, we try to explain what went wrong initially (see appendix A) and why this false conclusion still persists in its own alternate world of fundamentally flawed science that is unfortunately still widely accepted within the scientific community. It is even more widely accepted by the journalistic, environmental, policy-making, and general public communities.*

It is very difficult for people to accept something new when they were so comfortable with the old. Fortunately, at least some people are beginning to understand what is really happening to change our climate. For them, the confusing haze is beginning to clear.

One diffident sage pointed out that perhaps the appropriateness of the Nobel Prize being awarded to Al Gore and the IPCC back in

2007 might be compared with the presentation in 1948 of the Nobel Prize in chemistry to Paul Mueller[206], the Swiss developer and implementer of the then widely-acclaimed insecticide, DDT – which a decade-and-a-half later, was termed a "toxin" and "the most powerful environmental poison ever devised by humans" and shortly thereafter was banned from use by the United States and many other countries.

The Japanese Geophysical Union whose leadership had earlier enthusiastically endorsed the anthropogenic premise (and thus 100 percent of their membership had been counted as part of the "overwhelming majority" of believing scientists), took a poll of membership and found that 90 percent of their members were either unsure or believed the anthropogenic premise to be false. The "solar explanation" is beginning to be understood – or at least recognized – by many in the scientific community. Yet without a voice for the solar explanation, the highly biased public-opinion guidance message of the IPCC remains the only refrain being heard by the wider scientific, journalistic, environmental, policy-making, and general public communities.

For the first time, scientists studying the sun are explaining the internal dynamics of that formerly mysterious body and how these changing physical properties within the sun influence the changing climate here on earth in ways unimagined by atmospheric climate scientists. The atmospheric-oriented climate scientists had developed a simplistic but inaccurate version of the science that simply does not compare to the credibility of the more recently revealed explanations of the correct solar influenced science. Many of the applicable factors are summarized in the simplified matrix chart shown in graphic 16.2. (A more comprehensive summary of the earth's global warming/climate change paradigm is presented in appendix G.)

Factors Influencing Earth's Temperature and Climate.

Changing Factor.	Duration (yrs.).	Extent (ΔT).	Overall Impact.
1. Proximity of Earth to Sun.			
a. Base Milankovitch cycle.	100,000 ± years.	12°C ±	Major; Global.
b. Other Milankovitch cycles.	Tens of thousands of years.	2 to 5°C ±	Major; Global.
2. Cloudiness [Reflectivity of Solar heat radiation].			
a. Solar dual-magnetic wave cycle.	Multi-hundreds of years.	2 to 5°C±	Significant; Global.
b. Solar mass cluster transfers.	Half-century or so.	3 to 6°C±	Major; Global.
c. Particulates from all sources.	Continuous but variable.	2 to 5°C±	Significant; Global.
3. Total Solar Irradiance.	Up to a few decades.	1 to 3°C±	Relevant; Global.
4. Ocean and Atmosphere.	Continuous.		Minor; Mainly Regional.
5. Human Addition of Atmospheric CO_2.	One-time	Double CO_2 = ΔT of 0.6°C	Trivial; Global.

Overall relative impact is shown by type-size of numbered Changing Factor (1-5).

Graphic 16.2. **Factors Influencing Earth's Temperature and Climate** [207].

Thus, multiple mostly solar-influenced causes of changing global temperature account for all of the consequential changing earth temperature and climate experienced in the past as well as the present, and into the future – and over short-term, mid-term, and very-long-term time frames[208].

The Complete Picture

So at this point at the beginning of the third-decade of the twenty-first century, there is no legitimate theoretical evidence, no credible observational evidence, and no valid computer climate model evidence that the human use of fossil fuels has any consequential impact on global temperature or climate. And it has been shown that solar factors (for the variety of stated reasons) including the changes in cloudiness that are caused by solar processes (and thus earth's reflectivity of incoming solar radiation) are the principal causes of the global warming we are currently witnessing.

One interesting aspect of the process we are presently noticing is that it is quite likely that the earth has already started on the extremely slight and slow "orbitally" – and thus proximity-induced – temperature downswing. But the tiny temperature-change so induced because of the short time period involved, is being overridden by the much faster solar dual-magnetic wave cycle upswing in temperature. Thus, the earth's temperature will continue to rise until the current solar dual-magnetic wave cycle reaches its peak (most-likely within a few years) and then begin to drop-off. The current orbitally-induced downswing is so slow and so slight that it really is almost undetectable within a human lifetime – while the solar dual-magnetic wave effect becomes the measurable operative factor during this multiple-hundreds-of-years mid time frame period.

The Larger Perspective

Although not all the details have been worked out, the overall global warming and climate change picture is fundamentally clear:

If we look at earth's past 140-year, 360-year, 11,000-year and 420,000-year temperature histories we see a number of sustained global temperature changes clearly not influenced by either atmospheric greenhouse gases or human activities. The reasons for the changes in global temperature we see over these time frames appear to be changes within the sun and in the sun's relationship to the earth. Specifically:

- Varying radiant heat intensity from the sun's surface.
- Varying solar dual-magnetic wave cycles (several hundred to a thousand or so years) – which affect earth's cloud cover.
- Solar viscous-plasma cluster mass transfers that impart shorter term (half century or so) irregularity to the solar dual-magnetic wave cycle.
- Varying solar heat received by (as well as reflected away from) the earth that is controlled by earth's varying sustained cloud cover, which is regulated by the earth's varying magnetic shield – which, in turn, is strongly influenced by the changing solar magnetic field.
- Very slowly varying solar heat received by the earth caused by the changing distance between the earth and sun.
- We see no valid theoretical, observational, or computer climate model evidence of a consequential cause and effect relationship between atmospheric greenhouse gases and global temperature over any time frame.

Does Science Always Lead Us to The Truth?

As we've tried to show throughout this book, science leads us to the best answer available at the time, which is not necessarily the final truth.

Schopenhauer showed us how one person's "best notion" can lead millions to believe a precept which is not necessarily true even though it appeared that way at one time. The widely believed postulations discussed in Chapter 8: The world is flat; The sun rotates around the earth; Human sacrifice prevents natural disasters; Blood-letting cures human ailments; The Phlogiston Theory; and so-on, are all examples of not only a widely believed scientific wisdom, but an equally widely believed popular wisdom, as well. All looked pretty sound during their time, but ultimately all were found to be incorrect.

The question now is how long will it take the current scientific community and the public to recognize that they have accepted a false conclusion about the dominant role of greenhouse gases in global warming – and are operating under a climate change illusion?

Chapter 17
Conclusion

It isn't easy for people to change their whole way of thinking.

— Valley of the Tennessee (documentary film) 1944[209]

IMAGINE FOR A MOMENT how difficult it was when a few people began to tell the ancient scientific community that the earth was rotating around the sun – when "everyone" was sure that the earth was the center of the universe and could see with their own eyes that day-after-day the sun was rotating around the earth. To understand that the real mechanism was an earth rotating on its own axis which made it appear that the sun was rotating around the earth, required conceptual thinking. It also took the ability to cast-aside a conventional wisdom (a long-accepted belief). If, at that time, a poll of ancient scientists had been taken, would their consensus have made the overwhelming firmly-held geocentric belief that the earth was the center of the universe and the sun rotated around the earth, correct? Of course not. The parallel with the anthropogenic global warming consensus of today is striking.

There is no scientific basis for concluding that anthropogenic changes in atmospheric greenhouse gases have a consequential influence on global surface temperature or on climate. The widespread consensus that humans and their emission of greenhouse gases into the atmosphere is the principal cause of global warming is unfounded. *The cause of global warming is not increasing atmospheric greenhouse gases – it is changes within the sun, in ways not-previously-understood.*

Much of the educated world believes a fallacious conclusion that human beings and their emission of greenhouse gases is the prime cause of climate change when it is not.

Even a mere hint or vague suggestion to take a second look is met with animosity, derision, and ridicule by a combination of enraptured environmentalists, "chosen" climate scientists who have become captivated advocates of their own false supposition, and a hypnotized journalistic community. If ever there was a consortium able to lead the people of the world down another wrong scientific path such as those mentioned in chapter 8 (the world is flat, the phlogiston theory and so on) ... this is it.

There is a much more complex relationship between the earth and sun than originally thought by even the most qualified atmospheric climate scientists (or, for that matter, many solar scientists).

The scientific community as a whole, had a view of the role of the sun in the global warming paradigm as one of changing solar heat radiation along with a solar magnetic pole 11-year reversal cycle. They simply did not appreciate that a second longer term solar magnetic force, a multi-hundred-year cycle, influenced earth's cloud cover and the reflection of incoming solar heat. And that this cycle was a more important regulator of global surface temperature. Their unwarranted belief in the power of atmospheric greenhouse gases as a regulator of global temperature was so strong that they simply did not accept that changes in proximity between earth and sun were responsible for the very-long-term (orbital or paleo) global temperature changes that were being revealed by analysis of the deep ice-cores.

As the applicable processes – the shorter-term changes in solar irradiance; the mid-term magnetic and cloud heat reflectivity changes; and the much longer-term earth-sun proximity changes continue – we humans and our descendants are going to have to just hold on and try to do our best with the resulting consequential changes in climate, as we always have. It was probably a lot easier until fairly recently because there were so many fewer people (or "people-predecessors") on earth. Migration to a new unoccupied area was still feasible. There is considerable relatively recent evidence of this. In the western United States, when the area around Mesa Verde, in southwestern Colorado, dried-out some 735 years ago, the indigenous people living there simply migrated to an area where there was more water. The discovery and "opening" of the Western Hemisphere to Europeans some 500 years ago, provided an outlet for resettlement of the technologically advanced but "overcrowded" European continent. Such migrations became increasingly difficult as the population of the world began to exponentially increase at about that same time. Now nearly all the areas suitable for relocation are already occupied.

Recent human migrations are beginning to be into less densely populated but already occupied areas of greater prosperity. Sometimes this leads to friction between the current occupants and the new arrivals. And as the world struggles with its "growing population phenomenon" these migratory complications probably will increase. Demographers think the population of the world will level off at between 10 and 11 billion people, so the outlook should not be as ominous as would be indicated by a simple projection of the situation shown in graphic 17.1. Still, peacefully accommodating another three billion people into what many perceive to be an already overcrowded world will be a challenge for the people of the coming century or so.

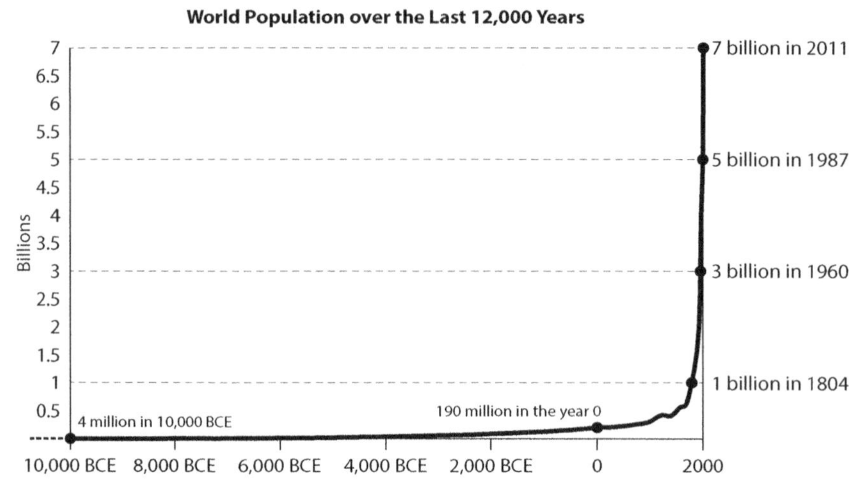

Graphic 17.1 **World Population Growth Over the Last 12,000 Years.** Max Roser, Hannah Ritchie, and Esteban Ortiz-Ospina [210]. [Time →].

It may take as long as 20,000 years or more to get from the elevated genial temperature plateau where we are now down to the next somewhat hostile "stadial valley" (graphic 15.4) and as long as 80,000 years to the depths of the next very hostile "deep-valley stadial glacial floor" following a slow, staccato, decline in temperature. The scary part is the realization of how severe the temperature difference between the "ceiling" and the "floor" of the long-term orbital cycle really is and the effects that could have on a much more massive world population (than the last time this part of the cycle occurred). People have already spread into now-temperate zones that will gradually become much colder and finally become uninhabitable. The more people the world has, the more difficult it will be for them to amicably adapt to these changing conditions. As the earth gets cooler and the "breadbasket" terrain shrinks, it will become increasingly painful to equitably share the resources available. As discussed above, we are getting a glimpse, a preview if you will, of this now – in the attempted migrations from

Southeast Asia and Africa to Europe and from Central America to North America – because there are too many people in those areas for the resources available within the culture that exists, and the region is simply unable to peacefully handle that many.

Back to the Current Climate

Recent observations by the European Organization for Nuclear Research (CERN) on a quantum physics level, show that even though the first derivative of the global surface temperature trend line has been positive for the last 20 years, it has decreased in value each month for those 20 years. This leads observant CERN scientists to believe that by the mid-2020s, the earth's surface temperature will begin to decline. It would appear that what the CERN scientists are noticing is the approaching end of the current 375-year global temperature rise caused by the solar dual-magnetic wave cycle as indicated by the central trend line in graphic 15.3 and the "final prong" (to the right) in graphic 15.2. It is indicative of the peaking of the solar dual-magnetic wave cycle and the "going-over-the-top" phase of the global temperature curve in both those graphics. Thus, as the earth's temperature swings over the peak of this cycle, it isn't going to get much warmer – and it surely is going to slowly start getting cooler within a relatively short time frame.

At this point, it will be difficult for those who have committed themselves so fully to atmospheric causes to look at the broader picture. They must begin to recognize the far greater importance of changing solar magnetic (and resulting cloudiness and reflectivity) effects. As well, they must accept the much longer-term earth-sun orbital proximity effects, bringing all of these to the forefront of influential factors, while discarding the greenhouse gas supposition as being of any consequence. This is indeed difficult for an atmospheric climatologist whose entire intellectual grasp of the subject of global warming has been focused on a fallacious perception of the role of greenhouse gases.

Virtually all of today's scientific papers that are based on the conventional anthropogenic interpretation of climate change, end with recommendations such as, *"This leads us to another aspect that must be researched and resolved…"*. And with unlimited grant funds available from multiple U.S. government agencies, the research just goes on and on. The human-caused global warming notion has become a moral force. It is so embedded in the psyche of the enraptured that it will be difficult for them to open their minds to an alternate, yet clearly more valid explanation of climate change – particularly one that is complex to understand and one which would diminish the need for further research in their specialty.

Anthropogenic thinking has become so pervasive that now even the children are being influenced. Some are suing their government[211], and others are conducting school strikes[212], all because "the adults" are not acting fast enough to stem the alleged "evil tide of human-caused climate change." And all based on a false premise of human culpability.

Is the false premise of anthropogenic global warming the grandest "big lie" ever promulgated on the human race (as many contrarians seem to believe)? Perhaps, but we prefer to look upon it as the result of an unfortunate scientific misinterpretation followed by several other human bumbles – and all as a result of humankind's continued efforts to understand the earth's natural surroundings with incomplete information.

Today, after they begin to accept the flaws in the "greenhouse gas supposition", much of the problem for atmospheric scientists is in appreciating the multiple causes of the sun's role. These combine to create a concatenation from the total heat radiation changes in the sun's surface; and the less well understood multi-hundred year solar dual-magnetic wave cycle (and the influence it has on earth's cloudiness and thus reflectivity of incoming solar heat radiation); to the "cluster transfers" of viscous-plasma mass between differentially rotating solar sections which cause multi-decade global temperature changes; and finally, to the changes of proximity of the sun to the earth. When linked together these are the principal causes of the short-term, mid-term, and long-term global temperature and climate changes we have seen in the past, are now seeing, and will see in the future.

The Elitist Deception

So, is anthropogenic global warming a hoax?

If the anthropogenic inner core of climate scientists told the public that something was true that they knew was not true – that would be a hoax. But in this case the inner core of climate scientists told the public that something was true that they knew might be true or might not be true, and in their judgement it was better for the public to believe it was true so they would take the actions necessary to curtail carbon dioxide emissions – just in case it later turned out to be true. These scientists honestly thought it was for the common good to deceive the public with an innocent "little white lie". What could be the harm? Thus, this little ruse might better be called an "elitist deception" because this small group thought they had superior judgement.

There are several problems with this kind of "we know better than you" thinking. One is that these scientists had no real-world conception of the problems associated with trying to convert the world to a wind and solar energy base (which was their thought) – or even if such a conversion was possible. Another problem is that these same scientists had no way of turning the story around if the other option (humans and their greenhouse gases are not the cause of global warming) started to turn out to be true. They had turned-on literally hundreds of thousands of scientists, politicians, journalists, and environmental activists who were operating as if their assumed option (the human-caused narrative) was gospel. It is very difficult to begin to turn around a movement of the magnitude they had constructed. And they didn't consider this possibility when they first proclaimed humans and greenhouse gases were responsible for global warming when they knew that it could be true or that it might not be true.

It makes it all the harder to accept the alternate (and what has turned out to be the valid) option when it emerges in bits and pieces as has happened in this instance. The tendency is to fight for what you originally thought and said. No one likes to be wrong. You've been winning – so fight on. A perfectly natural human instinct. So when do you stop fighting and begin listening to what's going on around you? All of this can result in a big scientific mess – like the one we're in now. It took about a century for the bogus phlogiston theory to run its course. Hopefully, the equally bogus anthropogenic global warming supposition (which has already been going on for nearly half that time) won't persist that long.

Well, that's a very small (but hopefully illuminating) part of the long sad tale of how nearly "everyone" – including most of the scientific community – has come to believe that the human use of fossil fuels is the principal cause of global warming when it isn't. A previously not recognized multi-hundred-year magnetic cycle within the sun affects the earth's magnetic shield and thus the amount of charged particles entering the earth's atmosphere. This influences earth's cloud cover and the amount of solar heat reflected away from the earth, which, in turn, regulates earth's surface temperature. This is a perfectly natural cycle over which humans have no control.

We can only hope that it isn't too late for the human-caused (anthropogenic) global warming false conclusion to be recognized not only as invalid but also why it is invalid, and how this came about more-or-less innocently enough. Climate scientists did the best they could with what they had when they had it. Now that they have more, let's hope they can swallow hard and begin to accept that things are not as simplistic as they previously had thought.

It has been a marvelous ride for many, many scientists and a remarkable never-ending source of funding. But now the time has come for all of us to recognize that changes within the sun, and the earth/sun relationship are the causes of global warming and climate change. The notion that humans putting greenhouse gases into the atmosphere is the cause of global warming and climate change is simply an extremely unfortunate climate change illusion.

The sun is our benefactor and the source of all we hold dear. Our atmosphere is precious and we must do all we can to protect it from pollution – but carbon dioxide is an atmospheric enricher – not a pollutant.

The earth has music for those who listen...

William Shakespeare[213]

Epilogue: Our Energy Future

Life without energy is short and brutal.

— Roy Spencer

IN 1949 THE AMERICAN geophysicist, M. King Hubbert showed us that as with any other buried natural resource at some point we will run out of fossil fuels[214]. Today, the world is about a third of the way through what he predicted would be about an overall 400-year world supply. Understandably, though, the "reserve supply" keeps getting supplemented and expanded as more and more reserves are discovered.

Historical Graph: The Epoch of Human Use of Fossil Fuels

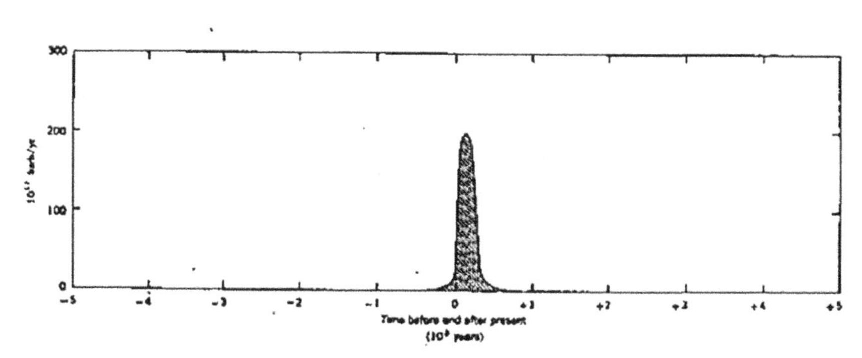

Graphic E.1: **Hubbert's 1949 Predicted Epoch of 400 years of Fossil-Fuel Exploitation in Human History During the Period from 5000 Years Ago to 5000 Years in the Future** [215]. *[Time →].*

Since this off-the-wall sketch there has been lots of thought and updating of the reserve of fossil fuels. In this book we have extended the end of the fossil fuel era out to between 400 years and 1000 years from now for a variety of reasons. An imperfect analogy might be made to that of Malthus and food supply being outstripped by population. Humans are quite ingenious. As the human population has increased, food supply yield per acre has also steadily increased. The yield for corn in the United States has increased ten-fold in the last 70 years. In addition to mechanization and irrigation, improvements have been introduced such as soil conditioning, (weed control and chemical fertilizers), plant biological modification (xenogeny [cross fertilization] and genetic modification). Through ingenuity and the advancement of civilization, human beings have been able to largely forestall the overall doom of Malthus's predictions (even though there are some gross pockets of inequality in this field).

Similar innovative advancements are already happening for fossil fuels. Improvements in auto gas mileage are phenomenal. Coal mines are being automated and formerly unextractable deposits can now be recovered. The fracking process has rejuvenated the oil and gas industries as formerly unusable tight shale deposits are now being widely exploited. Improvements in earth-moving equipment have made open-pit and surface mining more economical. Extraction of oil and gas has become economical from deeper and deeper water. Detection and definition of new fossil deposit techniques have vastly improved and will continue to do so (See textbox in chapter seven). The vast unquantified reserves of kerogen (a natural pre-petroleum-phase fossil mineral – sort of an exceedingly thick petroleum in consistency) were not considered by Hubbert.

The depletion of fossil fuel reserves will be much slower than

Hubbert shows in graphic E.1, as more and more deposits will continue to be discovered. But often they will be in less accessible locations. That means extraction will become more difficult. It's a matter of fossil fuels getting more expensive over the long term as they become more difficult to find and recover. Much of the petroleum low-hanging fruit has been plucked and we are moving into frontier and unconventional sources such as deep-water drilling and fracking.

This economic fact-of-life will mean that alternate forms of energy may become relatively less expensive within the next several hundred years. Although, as you have already seen in the case of wind and solar, the greater the penetration of baseload that these kinds of intermittently available forms of energy becomes, the greater their cost also becomes.

Probably the first big process change will be the need to convert natural gas (and coal) into the most favored fuel – a liquid substitute for gasoline, jet fuel, and diesel fuel for transportation purposes. Coal was widely converted into artificial liquid fuel by Germany during World War II and by South Africa when they were limited by anti-apartheid sanctions. SASOL, the South African descendant of that early effort is still quite active in the field of conversion from coal or gas to liquid fuel. Recently, Qatar, which has abundant natural gas reserves, originally in conjunction with SASOL and later Shell Oil, has constructed a massive plant for the full-scale conversion of natural gas to a liquid diesel-fuel product. Although this project is expensive (the cost of the plant escalated from an original estimate of five billion dollars to 24 billion), it still is projected to remain cost effective as long as the price of petroleum remains above $40 per barrel. And (with some trepidation) there are more of these gas-to-liquid (GTL) projects springing up around the world. Both natural gas and coal can continue to be the workhorses for electricity generation for emerging economies as well as developed ones until the phase-in of their replacement (probably controlled thermonuclear fusion, as discussed shortly). Then, after a very long transition, the remaining coal, petroleum, and natural gas, along with kerogen – can continue to be the source of hydrocarbon-based chemicals, fertilizers, lubricants, asphalts, and plastics as well as the raw material for liquid fuels.

However, we must anticipate the end of the fossil fuel era in perhaps 400 to 1000 years if for no other reason than cost (as demonstrated by the Qatar GTL project above). Humankind has become highly dependent on low cost fuel, just as we are on low cost shelter, food and clothing.

We have already addressed wind and solar as potential future base energy sources and found them lacking. But there are other "renewables" and "near-renewable" sources. The ocean (waves, currents, tidal differences, ocean temperature differential); geothermal; hydroelectric; nuclear fission; and biofuels. But after further examination, virtually all of these – with the possible exception of biofuels – are also lacking. For example, the ocean has a lot of energy available and its potential has been recognized. Over 1000 patents on various methods of harnessing the power of waves were granted in the United States during the twentieth century. But there are problems and challenges in utilizing ocean energy. Some of these are:

- Tethering (or anchoring) the generating device, is one stumbling block to successfully tapping the power of the sea. There is so much sand, sediment, and loosely compacted material on the sea floor that it's often difficult to get a good grip on anything substantial enough to securely fasten power generating equipment. And then, after considerable effort and expense for this foundation, the amount of energy produced by these mini-generators isn't very much.

- Marine organism fowling is another.
- Drifting seaweed and other undersea loose-material fowling.
- Corrosion in this hostile environment is severe and universal.
- Transmission of power to shore – problems with undersea landslides, whales, submarines, trawlers, curious scuba divers, small boat and ship anchors.
- Extreme weather events such as hurricanes, northeasters, winter storms, cyclones, and typhoons can cause havoc.
- Undersea maintenance is a bear.
- Local environmental problems include disrupted fish migration.
- Global environmental problems include slowed ocean currents that cause a disruption of traditional climate patterns, and restricted tidal flow that causes slowed rotation of the earth, the result of which makes for intriguing speculation.

Now one can understand why the forecasting agencies don't predict much energy from the ocean in the foreseeable future. Although nearly all these challenges could be overcome if it became absolutely necessary, it is unlikely that the ocean is going to be able to compete with an energy source such as the continued use of fossil fuels or the future use of controlled thermonuclear fusion.

Likewise, when carefully examined, other renewables have their own sets of restrictions. This leaves us with a very limited number of possibilities for a future base energy source.

Within about 500 years, humans, in addition to a very small (fringe amount) of renewables (hydroelectric, wind, solar, and geothermal), probably are going to be dependent on two basic forms of energy:

- Electricity generated by controlled thermonuclear fusion; or (much later) generated by elemental annihilation (both of which we will discuss shortly) will be used for all stationary uses (industrial, commercial, and domestic) as well as short-range transportation uses – electric cars, motorbikes, city trolleys, commuter trains, and short-haul trucks. Pre-charged quick change-out electric battery-packs (as is in-use today in China for some motorbikes) may become used throughout the world for mini-automobiles, as well.
- Hydrocarbon liquids will be used for transportation for long-range uses (interstate and international trucks, trains, ships, and particularly – aircraft). This liquid fuel can be produced either as a refined product from the remaining fossil fuels including coal and kerogen, or from cultivated plants (biofuels). Also, electric-powered reduction processes like the dissociation of water into hydrogen and oxygen and then the hydrogen could be used in fuel-cells (or the conversion of that hydrogen into a liquid by combining it with carbon monoxide or with nitrogen) could be in use. There will be a long term "phase-over" as the remaining natural hydrocarbons such as kerogen and residual coal, petroleum, and natural gas are reserved for lubricants, plastics, chemicals, fertilizers, asphalts, and other specialty uses.

It is remotely possible, if coal deposits significantly outlast natural gas and petroleum, that sophisticated steam engines with boilers fed by conveyorized crushed coal could return for ships (and possibly even trains).

Manufacturing process changes will influence the form of energy used. Once the world reaches "steel equilibrium" wherein nearly as many autos and motorbikes are scrapped each year as are built, for example, most steel will be produced from melted scrap and will use electric furnaces fueled by controlled thermonuclear fusion rather than the coal fired blast furnaces used today for "original" steel. Today some 70 percent of the steel produced in the U.S. is made in electric furnaces from recycled scrap rather than from iron ore.

As time passes, probably some of the kerogen family as well as residual coal and perhaps even charcoal from wood will supply the carbon for steel as well as for other products that require carbon.

Fusion versus Fission

The controlled thermonuclear *fusion* process for electric generation is quite different from, and inherently much safer than, the nuclear *fission* process that is currently used.

Controlled *fusion* involves the fusing together of light (or small) atoms, such as hydrogen, deuterium, and tritium and creating a larger (but still quite small) atom such as helium. At this light end of the element spectrum, fusing together of atoms gives off energy – in contrast to the breaking apart of heavy atoms such as uranium and plutonium at the other, heavier end of the periodic table which also gives off energy and is what we do today in the controlled nuclear reactors that we currently use in our nuclear power plants and nuclear ships. Thus, both breaking apart heavy atoms (today's technology – *fission*), and fusing together light atoms (tomorrow's technology – *fusion*) are potential energy sources.

We have substantially developed the "breaking apart at the heavy end" process (nuclear *fission*). Hundreds of power plants and ships (mostly submarines, icebreakers, and aircraft carriers) are powered by fission reactors in technologically advanced countries. Nuclear fission has some limitations, however. In the long term, it is dependent on the availability of a limited radioactive raw material (uranium ore). More immediately, to be kept under control during operation it requires a hot radioactive core that must be constantly cooled. This core gives off radiation which makes it potentially dangerous if, for any reason, the cooling substance (usually water) were to be cut off. Without being continuously cooled, this mass of radioactive material or "core" can heat up and lead to the dreaded "core meltdown" (perhaps with an accompanying large hydrogen bubble). This can result in the explosive release and dispersal of radioactive material, sometimes over the surrounding countryside.

A series of unfortunate accidents – some quite disastrous, such as Chernobyl in Ukraine Soviet Socialist Republic in 1986, and Fukushima Daiichi in Japan in 2011 – involved core meltdowns and the resultant release of radioactive material into the surrounding vicinity. These few tragic events have made for a bad public perception of the safety of nuclear fission and have probably ruled it out as a next-generation replacement for fossil fuels unless it becomes absolutely the only choice and the public becomes convinced that the benefits outweigh the risks.

There is some talk today about extending or optimizing nuclear fission (to make it more efficient) as a potential "next-generation" for the world's energy future. The problem with this route is that almost any significant efficiency improvement in the fission process involves increased risk. Several excursions in that direction involved the use of sodium instead of water as the primary loop coolant – which would have resulted in a more efficient fission reactor. But sodium proved to be far too corrosive of the pipes and the primary loop leaked. This became a lesson for what not to do. In the field of nuclear fission, you can't experiment with production-size reactors. You just don't have the luxury of trial and error in this field. One major strike and the whole field of nuclear fission might well be out. The public simply will not tolerate nuclear accidents that disperse widespread nuclear radiation, as we have seen from the few major nuclear accidents there have already been. As previously mentioned, these include: (a) Chernobyl in Ukraine SSR with an immediate 59 dead and thousands more killed as a result of the long, drawn-out burning radioactive core aftermath. Many of these later victims were conscripted to help ameliorate the radioactive damage, without being adequately protected and apparently without being made fully aware of the danger. As well, a half million people (spread out over hundreds of square miles surrounding the disaster area) were exposed to airborne radiation (fall-out). The tragic results of this ineptly handled debacle are still unfolding. (b) Fukushima Daiichi in Japan with possibly up to 16,500 dead (although with a massive tsunami also involved this is a calculated figure). This complex had four well thought out redundant safety systems designed to prevent a core meltdown. But no one had predicted the possibility a tsunami of the magnitude that came ashore, and thus all of these back-up systems proved to be inadequate. (c)Three Mile Island, in Pennsylvania (which was really quite minor compared to the other two with no dead or injured, but which caused no end of concern in the United States over the minor leaking of radioactive material). In any event, these three incidents caused enough concern about nuclear power to effectively dampen its growth and seemingly eliminate its prospects (at least for now) as a viable electric power replacement for fossil fuels.

But humans tend to forget. It may be that as time goes on and fossil fuels deplete, humanity will "forgive and forget" and welcome fission power back into the fold (particularly if controlled thermonuclear fusion has not stepped up to the plate).

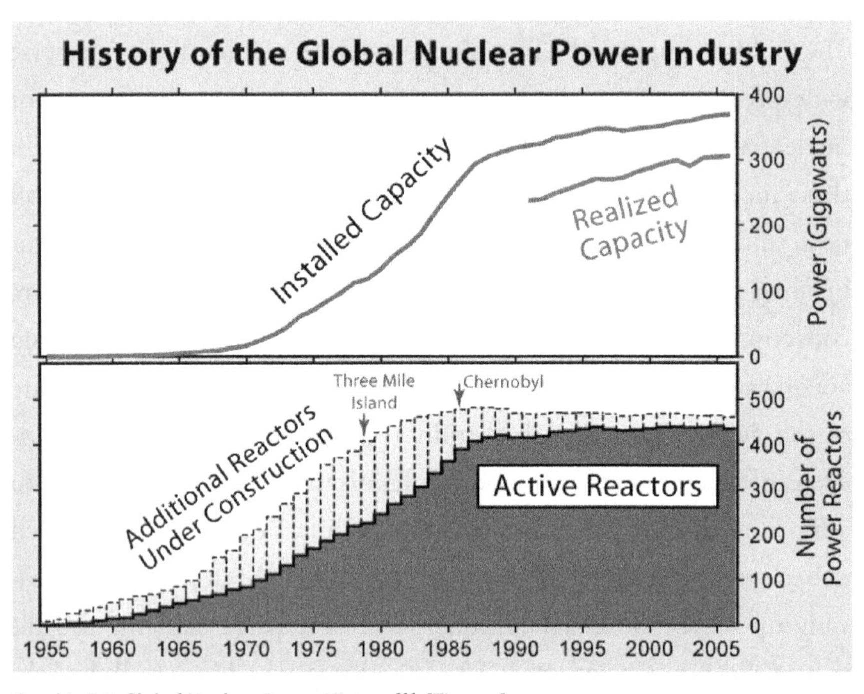

Graphic E.2: **Global Nuclear Power History**[216]. [Time →].

Molten Salt Reactors

(Including "Thorium" Reactors)

This technology has been around for more than a half century and has potential as a source for generating electric energy. Yet it has never been carried forward because it has so many unresolved issues that only a full scale "testing" phase might answer. So far, no major policy-making group has seen the justification in doing this. Molten Salt Reactors (MSRs) have recently received a resurgence of interest by people who understand the limitations of wind and solar but believe that there is a climate crisis as a result of the use of fossil fuels and greenhouse gases. To them, these MSRs are the solution that could be put in place in time to stave-off the catastrophe of continued fossil fuel use. Once they realize there is actually no climate crisis based on fossil fuel usage, then the pressure to prioritize MSRs will shrink in favor of an orderly succession from fossil fuels over a longer time frame to the apparent primary solution to our energy future – controlled thermonuclear fusion.

This is not meant to diminish the potential of the MSR concept, but to simply emphasize the need to clean-up our act with fossil fuels (that is, to clean the air) and continue to use them and allow their carbon dioxide to enrich the atmosphere until the manageable conversion to fusion electricity can be implemented in our cities and other regions – such as southeast Asia – where population density (and air pollution) are high. And then the world can move on to fusion as fossil fuels deplete.

The point is – let's not respond to false conclusions and climate change illusions. Let's respond naturally and economically to real needs and opportunities. These molten salt (thorium) reactors may well prove to be the "answer" if, as fossil fuels deplete, humankind simply cannot make controlled thermonuclear fusion into a practical, economical source of electric power for the future. The thorium reactor, assuming its current issues can be resolved, would almost certainly be a better answer than wind and solar.

Controlled Thermonuclear Fusion Is Safe

Materials at the light end of the periodic table – the hydrogen/helium end – suitable for the controlled thermonuclear fusion process, have an entirely different story than those at the heavy end.

First and foremost is safety. *Fusion* (fusing small atoms together) does not break the source elements up into radioactive pieces and particles as *fission* does – although, depending on technique used, some neutron particles may be given off. Even more important, however, is the fact that the *fusion* reactor does not involve a fixed critical mass of radioactive core material that must be externally and continuously cooled, as *fission* does, and therefore does not share the dangerous possibility of a core meltdown in its reaction process.

So basically, the controlled thermonuclear *fusion* process, while potentially much more powerful than *fission*, is inherently much safer than the nuclear *fission* process from a public safety viewpoint. It is vital that the understanding of the inherent safety of the *fusion* process be understood by the policy-makers of the world.

Today, the main raw materials for the European ITER[217] controlled thermonuclear fusion magnetically contained "tokomak" plasma project are deuterium and lithium. In theory if you fuse together four atoms of hydrogen you get one helium atom (and some energy) or better-yet, fuse together two atoms of deuterium and you get one helium atom (and some energy). In practice using today's technology the easiest way to achieve fusion is to force together one deuterium atom with one tritium atom which gives you a helium atom, some energy, and a leftover neutron, which harmlessly spins off. Tritium is very rare in nature and is currently bred from lithium. However you do it, the fusion reaction gives off lots of energy – but, at present, it is easier to do the deuterium plus tritium method and that is the ITER method that is currently on the brink of a successful breakeven reaction.

Deuterium is a component of seawater. About one in every several thousand molecules of seawater, whose molecular designation is H_2O, has a deuterium atom instead of a hydrogen atom combined with the oxygen atom or D_2O. To separate these deuterium/oxygen molecules from the mass of regular H_2O water requires processing lots of seawater, but it can be extracted by a fairly easy process and there is an almost unlimited supply of deuterium in the ocean. It is estimated that there is enough deuterium in the sea to supply the world's energy needs at thirty times the world's present rate of use for five billion years. If humankind extracted only one fifth of it, it would last a billion years. Lithium, however, is a limited resource and currently it is being used in medicines, greases, and in lithium/ion batteries – which are used in computers and cell phones, as well as in the growing market for electric automobile batteries. The lithium auto battery market should begin an exponential increase in lithium use as electric autos begin to become increasingly popular. That means the lithium battery market should already be in an exponential growth pattern when the lithium fusion reactor market begins its exponential increase in demand.

Lithium comes from fairly common rocks and brines located on five continents – with huge reserves in both South America and Australia. Perhaps the world mining industry can accommodate the "double-demand" envisioned for an extended period of time, but eventually an alternate to lithium would have to be found for one (or both) of the two major users – batteries and fusion reactors.

Scientists around the world have been aiming toward developing a process that would provide controlled thermonuclear fusion since about 1960 – and although the world's scientists have come close – they have not yet devised a fusion reaction that "breaks even" (i.e. that produces more power than it consumes). It is a frontier, cutting-edge technology with multiple directions to be explored. Some of these efforts result in costly failures, but fortunately motivated humans keep trying.

The co-operative, multi-nation, plasma ITER project has been successfully plodding along in Western Europe for more than a decade, and it looks like the current version of its tokomak plasma reactor will indeed break-even in the very near future. In fact, it is estimated this system will produce as much as ten times the energy it takes to drive it.

The following generation DEMO[218] project is also in the advance planning stages and is scheduled to follow ITER in an orderly succession. The United States is a "weak" participant in these international projects, supplying only some nine percent of the funding for ITER after an unenthusiastic on-again, off-again participation that has depended on fickle political winds. Fortunately, the leading force all along has come from more dedicated nations (mostly from the European Union). You might say that the U.S. comes in just behind South Korea and India in the effort. Our political policy conviction that the future of energy supply was in wind generators and solar panels caused a complete pull-out of U.S. participation from the ITER project when the political winds blew in that direction – only to very tentatively return since that time. And during the early years of the Trump administration again there has been talk of pulling out of the ITER fusion project because of the perception that fusion competes with coal. Actually, fusion will most likely come into its own as a successor to coal as depleting coal reserves are diverted to lubricants, chemicals and plastics. It will take a long time for controlled thermonuclear fusion to fully develop as a competitive production power source – and coal as well as natural gas will be needed in the interim. But we will need controlled thermonuclear fusion as the base energy source which will slowly replace coal and natural gas for electric-generating purposes and as coal and gas are diverted and converted to liquid fuel and other hydrocarbon uses such as lubricants, chemicals and plastics.

Today there is also debate about withdrawing our participation in the European fusion projects to finance our own. We should not be making a choice between these two alternatives – we should be doing both. This could be financed by withdrawing funds from subsidies for renewable (wind and solar) domestic projects and from research into the effects of human-caused global warming – all of which is a total waste of resources.

Renewables should be made to pay for themselves by their own economic viability without subsidy and with the recognition that wind and solar will require large capital expenditures to provide energy-storage capability to cover the mismatch between time of generation and time of use as penetration of baseload increases. The unestablished simplistic notions that motivated subsidy for wind and solar have been found to be fallacious. Unfortunately, at the present time much of our national (U.S.) financial resources in the energy sector are being frittered away on massive subsidies for renewable solar and wind projects that we already know can be a supplement to our existing energy base if we are ever desperate enough to need them.

Most of our former policy-makers apparently did not realize that wind and solar will become uneconomic when the penetration of baseload becomes large enough to make inroads into replacing the current fossil fuel base with them. Our current tremendous expenditure on wind and solar does nothing to stretch our scientific or technological competence, nor to improve our precision manufacturing capability. The subsidy money is mostly flowing out of this country into the hands of foreign entrepreneurs who are quick to take advantage of our government's largesse. We have politicians who are happy to get the crumbs (the foundations and installation work) for wind and solar while the meat and potatoes goes to outsiders.

This type of federal subsidy funding should be diverted to furthering the development of controlled thermonuclear fusion and other advanced particle-physics investigations just as soon as possible. Those who do achieve successful controlled thermonuclear fusion will reap rich rewards because it will become the long-term replacement of the earth's base energy supply into the future.

Currently, there are at least three privately (or semipublicly) financed projects trying to perfect small-scale versions of the magnetically confined plasma fusion process (One each in the UK, Canada, and the U.S.A.). These projects have financial backing in the range of $25 million to $500 million and are all using the concept of a modified tokomak barrel (or "cored apple") reactor. This method was discovered during the design and development of the ITER international fusion project underway in Europe (the one that includes minimal participation by the U.S.).

Undaunted by the two-billion-dollar failure in 2012 of the U.S. National Ignition Facility to achieve success using the laser-based Inertial Confinement Fusion (ICF) technique (wherein multiple laser beams are simultaneously and precisely focused on a series of tiny pellets of fusion fuel with the hope of a sustained fusion reaction) – a private firm is trying to raise funds for an attempt at a successful fusion reaction using the ICF method.

All of these entrepreneurial efforts are the beginning of a hoped for but perhaps premature transition of leadership in the field of controlled thermonuclear fusion from government to the private sector. The U.S. government will probably have to remain involved, at least during the developmental and early production phases, because these kinds of projects usually turn out to be so massive and so expensive that the private sector simply cannot possibly shoulder the financial burden while maintaining the high standards of safety and quality control required for enduring success in this field. And it is only fair for all of us to help reach the very elusive goal of successful fusion – since we will all benefit from success when it finally comes.

As far as the U.S. is concerned, the whole controlled thermonuclear fusion concept "suits" the U.S. paradigm. We have the innovative capability as well as the high-tech theoretical systems analysis and program management skills; the detail design competence; the versatile tight-tolerance heavy manufacturing proficiency; and the extremely high quality-control culture required for success in this field. The only things we seem to be lacking are the inspiration, the drive, and the will to make it happen. Do we really want to lead the way in the development of controlled thermonuclear fusion as the next-generation source of the energy needed to power the world, or do we want to fritter away our financial resources on mostly foreign-produced windmills and solar panels while we sit back and let others assume the technological, manufacturing, and economic leading role that we have performed for the world for the last 150 years?

We should be going all-out to develop and implement practical controlled thermonuclear fusion as an energy source for electric power. That way, as fusion power slowly phases in over many, many years, hydrocarbon deposits can slowly be diverted to provide for the growing need for raw material for chemicals, fertilizers, plastics, lubricants and asphalts as well as for the continuing need for liquid fuels for most long-range surface and air transportation purposes.

We must reiterate that controlled thermonuclear fusion is a completely different process than fission and does not present the potential for catastrophic failure as the Chernobyl or Fukushima nuclear fission reactors did. Fission reactors have a radioactive core that requires continuous cooling even after the reactor is turned off under emergency conditions, while fusion reactors have no such need.

Elemental Annihilation

Although humans are probably up to a decade away from successful controlled thermonuclear fusion, we are perhaps a hundred or more years away from elemental annihilation to produce energy.

Most people are familiar with the Einstein's famous equation: $e = mc^2$, which is a quantification of the relationship between matter and energy. It tells us that matter (m) can be converted into a lot of energy (e), since (c) is the speed of light (186,000 miles per second) squared. You can see there is an awful lot of energy waiting to be released in a tiny amount of matter.

The burning of fossil fuels does not convert any matter into energy – it simply releases the energy locked up in the binding together of atoms into a molecule.

In the *fission* process a small amount of matter is actually converted into energy as we break apart a very large atom into smaller pieces (as in our current uranium and plutonium nuclear power plants and nuclear-powered ships). In the *fusion* process, again, a small amount of matter is also converted into energy when we fuse together two (or more) of the smallest atoms to produce a larger one.

And after we perfect one method of controlled fusion, research will continue on the many other possibilities for fusion reactions that will also convert very small amounts of matter into energy in the hope of finding an even more cost-effective method.

But there is another process further along in this chain that potentially converts an even larger amount of matter into energy, and that is the process of *elemental annihilation*. Here, we (with great difficulty) produce a series of antimatter (negative) particles and stream them into similar particles of positive-matter, and thus annihilate both. This produces a tremendous amount of energy since most of that matter is converted into energy. The trick, of course, is to deftly create, control, and maneuver the antimatter with a lot less energy than is produced during the annihilation.

The Large Hadron Collider, sponsored by the European consortium CERN, is inadvertently delving into this arena (although that is not its prime purpose). Both the creation and handling of antimatter are tremendous challenges. But CERN is experimenting and this may lead not only to increased knowledge of particle physics (which is the basic CERN goal) but also to great ideas for the conversion of matter into energy. Don't expect a practical power source to come to fruition in this arena for some time, however. We should be making a strong effort to develop controlled thermonuclear fusion since it is well enough understood to become the energy source of the foreseeable future. Annihilation comes later, after we've learned a lot more about the very tiniest sub-particles in matter and what causes them to be positive or negative – as well as how to manipulate and control them.

There is much to be accomplished in the realms of both fusion and annihilation, and it involves exceedingly expensive facilities, equipment, experimentation, and development costs. Humankind must have a powerful economic capacity to be able to afford to delve significantly into these complex and very expensive areas. But it is imperative that the policy-makers of the world understand the need to do so.

Currently the United States is "wasting" its resources in two expensive arenas: First. we are subsidizing massive research into the effects of a phantom problem – that is, the environmental impact of projected human-caused global warming in order to try to influence the world to stop using fossil fuels so that the warming will slow-down. This is a complete waste of resources because the human use of fossil fuels is not a consequential cause of global warming – the sun is – and there is nothing we can or should do about it, because it is going to stop on its own in the very near future. We must learn to adapt until the earth starts to cool again, hopefully in just a few years.

The second arena where we are squandering resources is in subsidizing wind and solar production projects which do nothing to improve our technological capabilities in any way. Furthermore, as we have just seen, wind and solar are not the solution to our energy dilemma.

Both of these endeavors should be curtailed as soon as possible and the funding that is made available as a result should be diverted to projects that will bring us closer to a better place in the world. This involves advancement in two arenas:

First, we must solve our complex air pollution problems, many of which are directly related to the fossil fuels we are using and will continue to use for many years.

Second, we must recognize that over the long run we will require a replacement for our slowly depleting fossil fuels. That entails providing the technological and political support as well as the massive funding needed for all the steps it takes to bring controlled thermonuclear fusion on line.

Afterword

You can't stop what's coming...
— No Country for Old Men, (motion picture, 2007).

It is already obvious that humankind does not have the cooperative capability to take the steps necessary to curtail the growth of carbon dioxide in our atmosphere. It has been two decades since the *proclamation of anthropogenic cause* of January, 2001 that "confirmed" that we must stop using fossil fuels, and the whole world was acting on this ominous projection as if it were gospel. Yet the usage of fossil fuel throughout the world has increased and carbon dioxide has continued to be released into the atmosphere in ever increasing amounts. In fact, as of the beginning of the year 2020, the CO_2 content in the atmosphere was still growing just as fast as it has been since 1957, despite all the talk about the evils of carbon dioxide, all the wind generators and solar panels being installed and all the closing down of coal electric generating power plants in various "developed" countries including the U.S.A.. (See graphic AW.1).

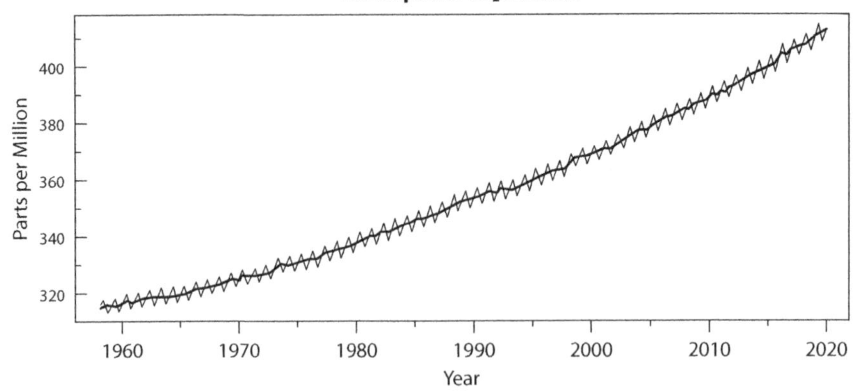

Graphic AW.1 **Atmospheric Carbon Dioxide Content 1957- 2020.** NOAA, Scripps Institution of Oceanography[219]. Note there is no reduction in the rate of growth from the year 2000 to the year 2020 despite massive "decarbonization" efforts. [time →]

With the switchover from coal to natural gas in many of our power plants, U.S. coal producers began to plan for export to willing users in Asia. This required West Coast exportation facilities. But the exportation of coal from the vast Powder River basin in Wyoming was blocked by environmental activists in Oregon (whose protests stopped the construction of the necessary export facilities).

Shortly thereafter, a massive new project to supply billions of tons of Galilee basin coal from northeastern Australia for a huge new power plant complex in India to supply electricity to Bangladesh, finally passed all approvals and was consummated in the summer of 2019. Obviously, fringe economies feel it is their turn to begin to enjoy the fruits of modern life, while the people of many advanced economies feel guilty and try to atone for past energy use "transgressions". Not surprisingly, the much more populous fringe economies are winning this psychological struggle. So the march to a higher atmospheric carbon dioxide content continues unabated despite still being aggressively opposed by anthropogenic environmental activists. If the doomsayers had been correct, and there was actually a real problem with carbon dioxide, the outlook for our world would by now be grim indeed.

Fortunately, all that is behind us. Carbon dioxide has no consequential detrimental effect on global temperature. In fact, as we have seen, CO_2, is an atmospheric enricher. Your grandchildren now have their future back (from a global warming perspective). All those climate change horror stories are nothing but illusions based on fallacious assumptions. Greenhouse gases are not the principal cause of global warming – and changing solar magnetism and accompanying earth cloud factors are. And since the current cycle of this kind of natural warming is going to peak very soon and turn into a slow cooling trend of its own accord, human beings are just going to have to learn to accept the little change the sun will bring us in the next few months or years – and wait for the next sustained global temperature turn which is predicted to start within less than a decade (and actually may have already started).

The COVID-19 Pandemic Effect

The overall effect of the 2020 COVID-19 pandemic is unfolding as this book goes to press (June of 2020), but with the imposition of "social distancing" measures as a means of controlling the spread of the virus until the mass availability of an effective vaccine (predicted for early 2021) there will be at least a temporary reduction of energy use in the U.S. and around the world, and thus a reduction of the amount of carbon dioxide being emitted into the atmosphere.

During the first quarter of 2020, the *transportation* and *travel* segments of the "developed world economy" declined significantly compared to what they had been at the end of 2019[220]. This has led to a noticeably consequential reduction in the use of liquid fossil fuels – resulting in a tremendous world oversupply (until production is able to get back in balance with demand). Likewise, with closed schools, offices, restaurants, factories, stadiums, and so on, there has been a reduction in other forms of fossil fuel use as well. Thus, the pandemic might do what human cooperation was unable to do – reduce carbon dioxide emissions into the atmosphere – at least slightly and at least temporarily.

But no matter how much or how little greenhouse gas is emitted into the atmosphere, it will not influence global temperature. In fact, global surface temperature will soon start to slowly decline in accordance with the principal sustained mid-term regulator of global surface temperature – changes within the sun, most notably its dual-magnetic wave phase progression cycle.

If you are an anthropogenic global warming environmental activist, and you can open your mind to what has been revealed about the science of climate change over the last few years – you can begin to stop worrying about human-caused global warming; and perhaps find a new, more suitable environmental cause for your concern. How about the destruction of forest land? Or the mass extinction of various animal species?

The End

Let us cultivate our garden.

— Voltaire, 1759[221]

Appendices

Appendix A: Water Vapor Saturation

As pointed out in chapter 15, water vapor is present in most of the atmosphere. Clouds are formed from droplets of condensed water vapor. Each droplet in a cloud needs a nucleus (particle) upon which water vapor can condense before a cloud can form. The availability of these droplet-condensing nuclei can become the limiting factor in the formation of a cloud. The extent of cloudiness affects global surface temperature because, among other things, it affects the amount of incoming solar radiation that is reflected away from the earth.

But when considering climate change, we are also concerned with the earth's outgoing infrared radiation trying to escape the earth. Specifically, we are concerned about the earth's infrared radiation that is trying to escape through the so-called "open" (cloud free) sky. In this area, there is water vapor present which is able to absorb some of the earth's outgoing infrared heat and there is also carbon dioxide which is able to absorb a small amount of outgoing infrared heat within a very limited infrared absorption band. The water vapor within this "open sky" atmosphere has a molecular spectroscopic feature called the "atmospheric window" discovered by Hettner and Simpson that we mentioned earlier. When we are interested in carbon dioxide and the role it plays in absorbing outgoing heat from the earth, we must remember that water vapor will be present absorbing heat emanating outwardly from the earth in the infrared absorption band zone where the spectroscopic infrared absorption bands of the two greenhouse gases water vapor and CO_2 are overlapped. And in this overlapped zone, when you add a little, or even more CO_2 it doesn't influence global temperature because in that zone the water vapor is already absorbing all the earth's outgoing infrared radiation. Unfortunately, inner core climate scientists split the heat absorbing effect 50/50 to each water vapor and CO_2 in this overlapped zone for their calculations. And this 50 percent (to each) split was used to calculate the influence of carbon dioxide on global temperature, resulting in the calculation of a 1°C rise for a doubling of atmospheric CO_2 (and which was further amplified by water vapor feedback into a 3°C rise in global surface temperature). But for the use of determining global temperature increase as a result of rising CO_2 in the atmosphere, it is necessary to give virtually all of the effect in that overlapped zone to water vapor. This is the most difficult point for people – even scientists – to comprehend. It comes down to this: although on the surface it would seem appropriate to mathematically split the effect evenly between water vapor and carbon dioxide (and no doubt this is the basis for climate scientists doing so, and for that 50/50 split passing peer review, and for other unassociated scientists quietly acquiescing, "*hmmm, yes, that seems logical*"), it is not correct in this instance. Since we are trying to determine the influence of carbon dioxide on a warming globe, we must look at it from that perspective. *If we removed all the carbon dioxide from the overlapped zone, would there be any effect on global temperature? No, because water vapor would still be absorbing all the outgoing heat in that overlapped zone. Likewise, if we added double the amount of carbon dioxide into that zone would there be any effect on global temperature? Again, no, because water vapor is already absorbing all the outgoing heat in that zone. So in that overlapped zone, nearly all of the greenhouse effect must go to water vapor when we are trying to determine the global surface temperature effect of a change in the carbon dioxide content in the atmosphere.* And since the overlapped zone is such a large part of the "plugged" area within the atmospheric window, the error is very large and results in an incorrect conclusion, elevating the influence of CO_2 from trivial to consequential (see molecular spectroscopy section below). In fairness to climate scientists they were not alone in misunderstanding this very vital point. As this spectroscopic approach has been "peer-reviewed" by other scientists of the very highest reputation, this fundamental error has been missed. And it has been carried through the scientific discourse of successive generations of scientific papers well into the twenty first-century[222]. Unfortunately repeating it doesn't make it correct. It was a simple oversight that has been mindlessly carried through from one generation of scientific papers to the next. Yet it has a tremendous influence on the world's perception of the importance of the effect of atmospheric greenhouse gases on climate by grossly underestimating the effect of water vapor while enormously overestimating the influence of atmospheric carbon dioxide on global surface temperature.

This simple oversight is the primary cause of literally billions of people believing that human emission of carbon dioxide into the atmosphere is the principal cause of global warming when it is not.

The Molecular Spectroscopy

Since this is probably the least understood but most important technical point in the controversy over humankind's role in global warming, we give an expanded explanation here:

The sun radiates heat toward the earth at the low wavelength end of the infrared spectrum. The earth, being a much cooler body than the sun, radiates heat back out at a slightly higher infrared wavelength range. It is this outgoing infrared radiation that is partially trapped by greenhouse gases (principally water vapor and, to a much lesser extent, CO_2 and other trace greenhouse gases) as well as by clouds, which causes the earth to be about 33°C warmer than it would be without the earth's atmosphere (this is often called the greenhouse effect).

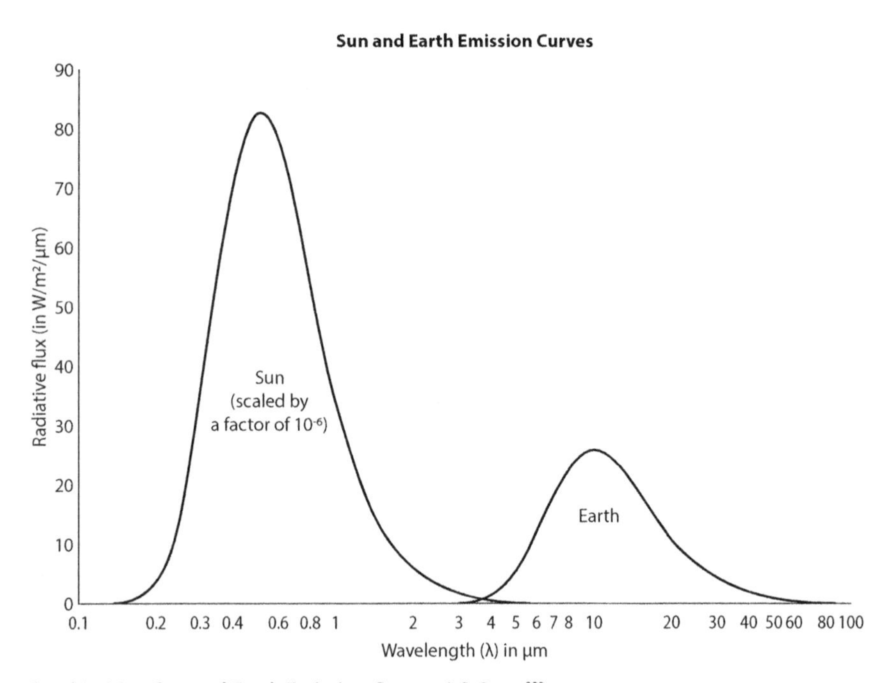

Graphic AA.1. **Sun and Earth Emission Curves,** J.C. Baez,[223]

Note that the bell curve of earth's outgoing infrared radiation starts at about 4 μm (microns) and runs to roughly 50 μm and peaks at about 10 μm. Any "greenhouse gas" in the atmosphere with heat-absorbing qualities within the range of about 4 μm to about 50 μm can thus influence the greenhouse effect. (Although the closer to 10 μm, the greater the influence).

Next, we need to look at Graphic AA.2, Atmospheric Absorption Bands and note which bands will be affected by the earth's outgoing infrared heat. For example, the methane "blip" between 3 and 4 um would have no effect because it is outside the band where the earth is emitting heat while a similar "blip" of nitrous oxide between 5 and 6 um would have a very small influence since it appears to not have much overlap with water vapor – but it is in the very low part of the bell curve of outgoing heat from the earth (graphic AA.1).

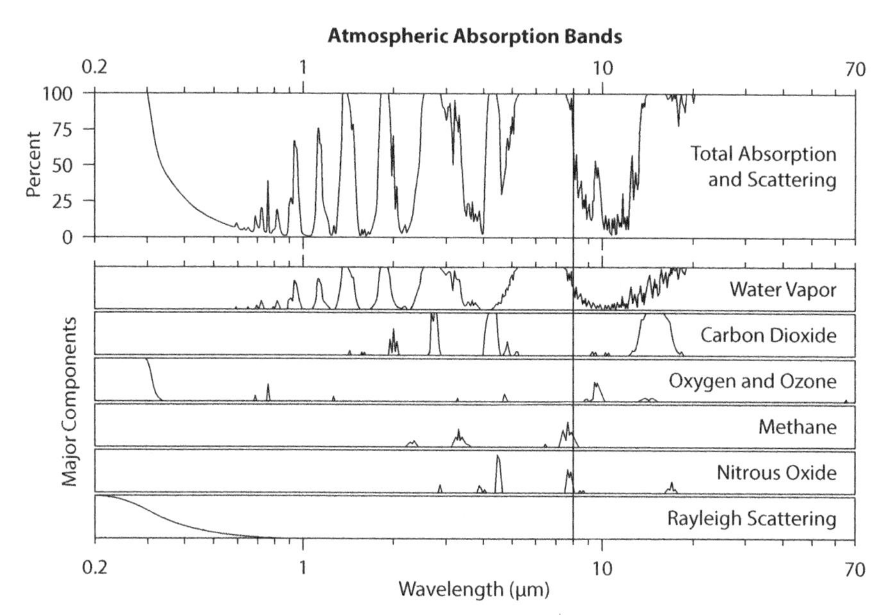

Graphic AA.2: **Atmospheric Infrared Absorption Bands.** R.A. Rhode. (with indicator line added)[224].

The methane and nitrous oxide blips between 7 and 8μm would be quite consequential if they were nor overlapped by water vapor (which they are) so the water vapor overlap cancels out nearly all of their effect. Thus, it is important to note the overlap with water vapor as well as the earth's heat emitting spectrum to determine the full effect of a trace greenhouse gas. Another significant point that many researchers have missed is that the HITRAN database for water vapor was revised in 2004 and the "atmospheric window" (shown here from about 7 to 20 μm) was reduced in size by about 25 percent from the values depicted in a popular encyclopedia published in 1992[225]. This change in the atmospheric window meant that the only applicable blip for methane now became nearly completely overlapped by water vapor. It is possible that the older information was used to draw the incorrect conclusion that methane had a consequential effect on global surface temperature when it didn't.

Next is shown the infrared (IR) absorption bands of both water vapor and CO_2 for the crucial area where earth's infrared heat is trying to escape but is being partially blocked by overlapped atmospheric gases (graphic AA.3). This is a crucial picture to understand – particularly the overlapped area at about 15 μm. The CO_2 "plug" from around 14 to 18 μm has some effect. However, water vapor significantly overlaps this CO_2 plug, and is already absorbing outgoing heat radiation. Thus, some heat "escapes" through the "open" area and CO_2 "blocks" some of the opening and thus does have an effect. But what about the overlapped area where both CO_2 and water vapor block the same area?

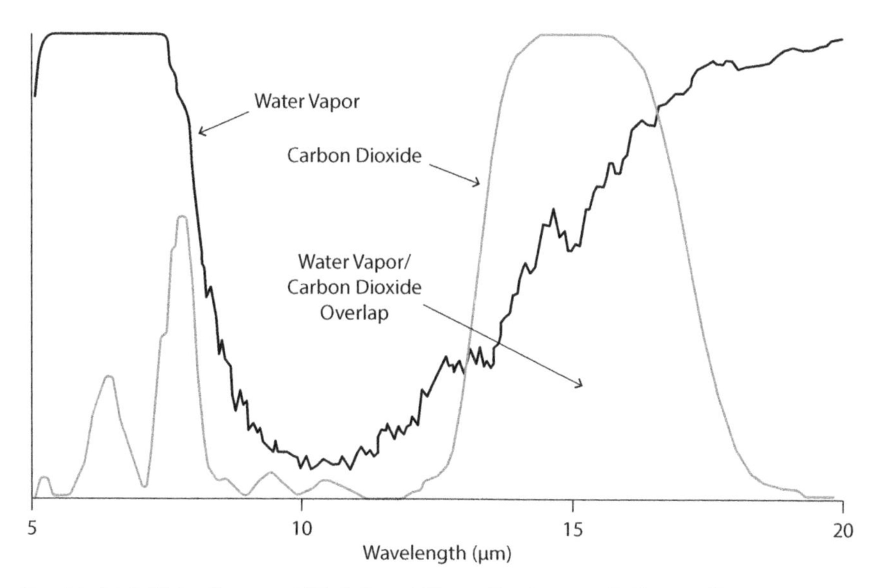

Graphic AA.3: **Water Vapor and CO_2 Infrared Absorption Spectra.** R. Simmon[226].

How do we treat this overlapped area? This chart is based on a one molecule to one molecule analysis. Yet water vapor's concentration is at least 12 to 15 times greater than CO_2's so should we give perhaps 92 percent of the weight of the overlapped area to water vapor and 8 percent to CO_2? Inner-core climate scientists gave it 50 percent to 50 percent. This was because they did their analysis of each gas in isolation (meaning that they compared one molecule of CO_2 to one molecule of water vapor while each was isolated from other gases) and thus half the overlapped zone was credited to each. Another researcher used 92 percent to 8 percent, basing his calculations on the relative concentration of the two gases. But the correct answer is nearly 100 percent to water vapor (as noted above in the saturation discussion) because we are trying to determine the effect of a changing concentration of CO_2 on the system. With 12 or more times as many molecules of water vapor already there totally saturating that overlapped area and absorbing all the outgoing infrared radiation in that overlapped area, there simply would be no effect by adding a CO_2 molecule (or two CO_2 molecules). *For that overlapped zone it's like covering a window with 12 (or many more) nearly-opaque black blinds – all the light is blocked – and the addition of a dark blue nearly-opaque blind (or two dark blue nearly-opaque blinds) will have no effect on how much light comes through because it was all stopped by the sum of the 12 or more black nearly-opaque blinds.* The inner-core and the IPCC, as well as the myriad computer climate modelers, continue to treat this area with a 50/50 split thus attributing an effect to CO_2 that it simply does not have. And as you now see this is one of the prime theoretical misinterpretations that has a heavy influence on the ultimate outcome of why so many scientists (and the general public) think that humankind is responsible for global warming when it is not.

Correcting this interpretive error (combined with updating of the water vapor infrared absorption band configuration in the HITRAN molecular spectroscopic database) causes the effect of doubling carbon dioxide content in the atmosphere to drop from 3°C (a consequential figure) to a five-times lower 0.6°C (a trivial figure)[227].

Appendix B: Solar Forcing versus Greenhouse Gas Forcing

The final link in the inner-core's logic chain disintegrated when it was discovered that there was indeed a very logical cause of a warming earth other than increasing atmospheric greenhouse gases: *global surface temperature changes caused by magnetic (rather than just heat radiating) properties within the sun and also by the sun's proximity to the earth.* The inner-core had tried to demonstrate that changes in solar irradiance were simply not powerful enough to cause the changes in global temperature we were seeing. A series of charts and graphs had been published showing this. Graphic AB.1 showing massive greenhouse gas forcing and very little solar forcing is an example. These charts stem from the time when the climate scientific community thought that 50 percent of the radiative forcing could be allocated to carbon dioxide in the infrared band where it overlapped water vapor – thus giving it a boost it didn't deserve.

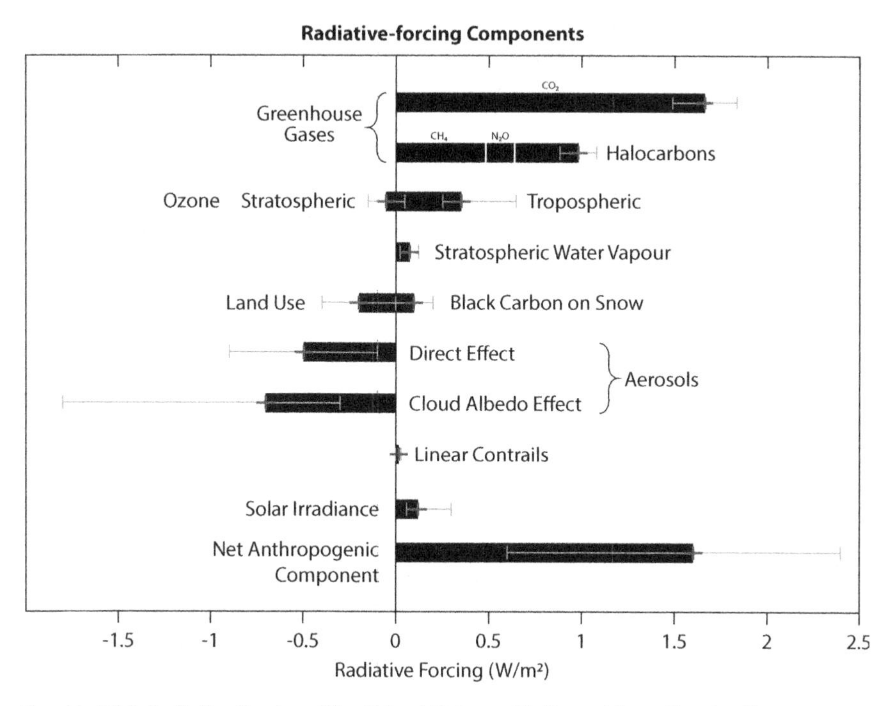

Graphic AB.1: **Radiative Forcing.** After Leland McInnes [228]. (Repeat from Chapter 1).

These charts stem from earlier ones with essentially the same information that was passed down from chart generation to chart generation with minor refinements. Apparently, each new generation was approved by peer review based on the acceptance of the earlier generation and without any critical reanalysis of the essential base information along the way.

The next two charts are distilled down to the salient points of discussion and only show the influence of carbon dioxide, methane, and nitrous oxide (greenhouse gas bar) compared with solar forcing. Other qualities, both plus and minus, are excluded because they are distractive and not germane to the central discussion (even though they do presumably have influence).

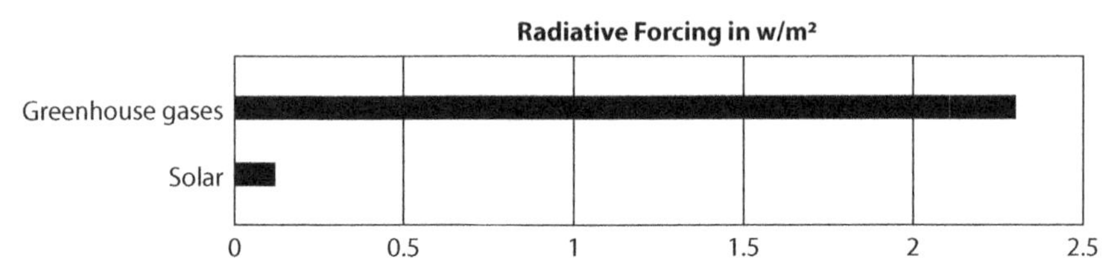

Graphic AB.2: **Radiative Forcing** [229]. Chart derived from graphic AB.1 by the anthropogenic inner-core but with inapplicable information stripped. Data, IPCC; Chart E.R. Pryor.

Note that this depicts major influence on global warming by greenhouse gases and very little by the sun, a main tenet of the anthropogenic position.

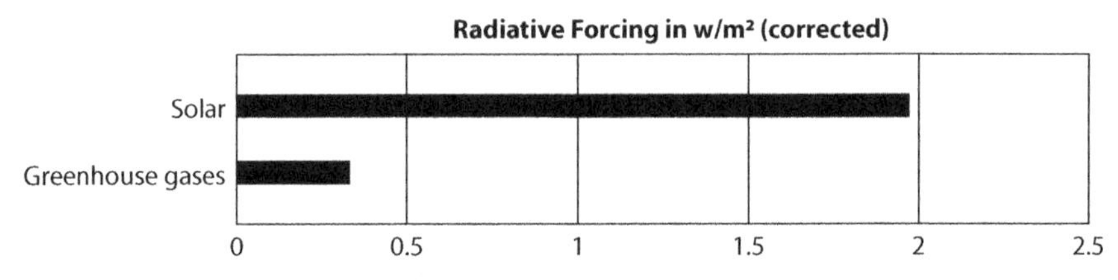

Graphic AB.3: **Radiative Forcing**[230]. Chart updated for reduction of inappropriate greenhouse gas forcing and addition of appropriate solar forcing (see detailed explanation).

Remember from the earlier discussion about the overriding role of water vapor's absorption band, that the water vapor band nearly completely overlaps both the methane and nitrous oxide bands, and substantially partially overlaps the CO_2 band. Thus the greenhouse gas influence is materially reduced from what the charts produced by the anthropogenic scientists depicted. Corrections (reductions) to the greenhouse gas forcing bar shown in graghic AB.2 (anthropogenic position) include: overlapping absorption bands between water vapor and each gas changes the effect of: (a) methane from 0.48 w/m^2 to 0.048 (one tenth of the IPCC figure); (b) nitrous oxide (N_2O) from 0.16 w/m^2 to 0.016 (also one tenth the IPCC figure) – both these reductions are because of the overlap between their IR absorption bands and water vapor's; and (c) CO_2 from 1.66 w/m^2 to 0.33 w/m^2. This last figure is arrived at by using the mathmatics from the Harde paper[231]. You will recall that both nitrous oxide and methane's infrared absorption bands were nearly completely overlapped by water vapor's and thus have almost no effect on global temperature – so they are reduced to only 10 percent of the forcing effect shown by the anthropogenic calculations. The same is partially true for carbon dioxide as explained earlier and its effect is appropriately substantially reduced. The influence of the sum of the greenhouse gases CO_2 + N_2O + methane, has thus been reduced to 0.394w/m^2 from 2.3w/m^2 or some 5.8 times lower than the anthropogenic position figures.

Likewise you will remember that the anthropogenic scientists only saw very minor influence by the sun. This stems mainly from a failure to understand that there are multiple solar forces when they saw only one.

In our new chart, the total forcing (greenhouse gas plus solar) becomes 2.42 w/m^2 (the total figure used by the IPCC). Then the new bars would be: 0.394 w/m^2 for greenhouse gases (as calculated above) and 2.026 w/m^2 for solar. Thus, the solar forcing bar increases from 0.12 w/m^2 (shown in graphic AB.2 [anthropogenic-position] chart) to 2.026 w/m^2 (nearly 17 times greater) shown in Graphic

AB.3 (corrected) chart – based on adding the influence of differences in earth's changing cloud cover which, in turn, is based on the changing magnetic shield surrounding the earth as a result of the solar dual-magnetic wave cycle. These earth magnetic shield (combination of earth and sun magnetic fields) changes, alter the pattern and quantity of solar wind ionic particles and cosmic rays entering the earth's atmosphere that create particle showers – and thus the formation of clouds. This, in turn, changes the earth's reflectivity to incoming solar heat radiation and thus strongly influences global temperature.

We can see by comparing graphic AB.3 to AB.2 that there is a complete reversal in influence by the sun and by greenhouse gases from what the anthropogenic scientists depicted in their chain of "radiative forcing" charts because their logic path was developed while they were unaware of their incorrect allotment of greenhouse effect between water vapor and CO_2 (the erroneous 50/50 split in the overlapped zone). They also were unaware of the proper multiple signficant influencing factors of the sun over and above solar heat radiation – particularly the solar dual-magnetic wave phase progression cycle.

Appendix C: The "Rescue-Teams" Efforts

As mentioned in Chapter 11, at least three "rescue teams" were dispatched to try to confirm that somehow increasing atmospheric carbon dioxide was causing the global warming noticed in the paleo (or orbital) long term Antarctic Vostok ice-core data rather than the observation by Barkov that the earth began to warm hundreds of years before the atmospheric carbon dioxide began to rise, and therefore some other factor (such as the sun) was causing both the land and ocean to warm and the climate to change.

One team was finally able to reduce the lag between when the atmosphere began to warm and when the carbon dioxide content of the atmosphere began to rise (originally postulated by Barkov to be from about 400 to 800 or so years) down to about 200 or more years[232]. The team had discovered an apparently valid point: the air diffused upward as the compacting snow turned to ice thus making the air bubbles younger than the ice surrounding them. This revelation removed some of the lag of rising atmospheric CO_2 from the leading warming global temperature – but no matter how hard they tried, not all of it. Rising CO_2 still lagged by about 200 or more years. This still demonstrated that the rising carbon dioxide was a consequence of the warming earth rather than a cause of it. The best rebuttal that the rescue team could concoct was to declare 200 years a "mere blink of an eye" and move on without explaining the 200-year lag.

Another rescue team came up with an interpretation favorable to the anthropogenic cause: they acknowledged that the pacemaker was the 100,000-year Milankovitch cycle of a nearing sun, and that it initiated the rising global temperature at the end of each glacial age cycle. Then, according to their "rescue scenario", the rising atmospheric CO_2 that was being off-gassed by the warming ocean 200 or so years later, had caused the temperature amplification that led to the next genial age. It should have been evident, however, that if there were an amplifier it would be the much more abundant water vapor which would act rapidly – yet would not be detectable in an ice-core analysis because the water vapor would simply freeze into the mass of snow being compacted into ice. And that the rising CO_2 was nothing more than a 200-year-later noncontributing "tracer". Thus, apparently the reason the earth continued to warm was not because of rising atmospheric CO_2 but because the earth and sun continued to move closer to each other as the basic Milankovitch cycle progressed.

Any feedback amplification would have been due to the much more prevalent and faster evaporating water vapor rather than CO_2, but it would not be obvious because the water would have frozen into the ice and not remained as a vapor in the bubble.

A third rescue team made a creative effort to try to show that what happens in the arctic regions doesn't apply to the rest of the world[233] and therefore it is possible that rising atmospheric carbon dioxide was indeed the cause of rising temperature. They used a postulated computer modeled bi-polar heat seesaw to explain how temperature changes in the Antarctic could be one thing while temperature changes in the rest of the world were something else. And it certainly deserved an A+ for creativity in a world of possibilities.

However, no matter how you look at it, the fundamental operational function is that the carbon dioxide content of the air during that era was almost certainly the *result* of off-gassing from a warming ocean, and not the *cause* of that warming.

As in the case of the hockey stick, this whole concept is simply an exercise in futility. Clearly other forces – most notably the changing distance between the sun and earth is the principal cause of changing global temperatures of this magnitude and over these long-term time frames. Attempts to portray changing greenhouse gas content as the cause of long-term global temperature changes can certainly be expected from scientists whose focus is to attempt to illustrate that the driving force of climate change is rising greenhouse gases. But we have illustrated that rising greenhouse gases have only a trivial effect on climate change. Thus, they cannot be responsible for the global temperature changes that occurred. It is more productive to look elsewhere for the cause of long-term climate change. Lo and behold – there it is: Solar proximity to the earth caused by the

Milankovitch cycles.

Apparently, these "last ditch" teams were not aware of the growing evidence of invalidation of the physics supporting the greenhouse gas thesis as an important factor in global warming during the time of their investigation. And, unfortunately, they had no explanation of the earlier periodic warmer temperature periods shown in graphic 10.3 and AG.2. Thus, their efforts were doomed from the start.

Appendix D: Ocean Acidification

Ocean acidification has been termed the "evil twin" to "global warming" by some environmentalists. There may be some merit to that phrase depending on what time frame you are concerned with, or when it is applied to the potential for disturbance of the chemical balance of the ocean at a faster rate than scientists would expect from a so-called natural process. Any redistribution of species or adaptation required by sea life for increasing CO_2 in the marine environment may be a cause for alarm by environmentalists who want to maintain things as they know them to be now and are afraid that any change would be a change for the worse. But most scientists recognize that life will very slowly adapt or evolve to the new slightly less alkaline conditions caused by humankind's emissions of CO_2 into the atmosphere – some of which, no doubt, will be absorbed by the ocean. The transfer of CO_2 back and forth between the ocean and the atmosphere is a complex process that depends on: (a) the concentration of CO_2 in, and the temperature of, both the atmosphere and the ocean; (b) ocean temperature and circulation patterns at different depths; (c) atmospheric temperature and pressure; and (d) various biological processes. All of these have an effect.

Most of this subject is too far from our central theme for a detailed discussion here, but the following factors are important:

There is no danger of the ocean becoming acidic. The ocean is still quite alkaline (basic) although it is trending in the direction toward neutral as some atmospheric CO_2 is absorbed by it. If it should ever get to the alkaline/acid neutral equilibrium point (7.0 pH), however, the trillions of tons of calcium carbonate sands, sediments, and long-ago discarded empty sea shells in the ocean will provide a buffer to neutralize any tendency toward actual acidification – and thus the ocean will never become acidified even if we deplete the entire global reserves of fossil fuels, emit the resulting carbon dioxide into the atmosphere, and a significant portion of it is absorbed by the ocean.

One main worry today, is the effect of the trend toward ocean acidity on coral reefs[234]. The mere "reduction of alkalinity" currently going on may have an effect on the accretion versus dissolution of some calcium deposits in the ocean, such as coral reefs. Recent studies cast some doubt on the widely assumed "truth" that this is a consequential long-term factor, however. Both laboratory experiments and field studies of coral reef skeletal growth have found the process to be quite complex and perhaps ocean "acidification" does not have the straightforward long-term effect that it is believed to have by many ocean biology scientists[235] (and the public).

Currently, possibly 25 percent of humankind's atmospherically released CO_2 is absorbed by the ocean (while perhaps about 25 percent is absorbed by terrestrial vegetation and somewhere in the range of 50 percent remains in the atmosphere to enrich it for future generations of terrestrial and marine plant life.) All this holds for the current atmosphere/ocean balance, which may drastically change as the solar magnetic/cloudiness/reflectivity forces (over the mid-term) or the sun separating from the earth (over the long-term) cause the world to become cooler. Thus, it is fruitless to worry about the future chemical balance of the ocean if we assume today's conditions when we know those conditions are going to naturally change within various geologic time frames.

Also remember that the vast quantities of oceanic phytoplankton – as well as Sargasso and shallow water sea grasses, algae, and other underwater vegetation – require carbon dioxide as well as sunlight for their sustenance, just as terrestrial vegetation does. Thus, increased carbon dioxide that is dissolved in the ocean is an enricher for them. Remember that this marine vegetation provides not only food and shelter for marine life, but also much of the oxygen we terrestrial animals breathe – perhaps as much as 70 percent. From that perspective, the ocean can accept much more CO_2, just as the atmosphere can. Since it doesn't directly apply to the causes of global warming, we'll leave this complex and highly debatable subject to the specialists. If they balance their outlook, they probably will find that whatever negative effects there are on oceanic life as a result of an increased reduction of alkalinity ("ocean acidification") will be counterbalanced by the positive effects of increased carbon dioxide in the sea water. And as we stated, the whole picture will slowly change as the ocean cools in response to the natural cooling that is coming.

Appendix E: Interconnecting Electrical Grids for Wind and Solar

Some planners think the mismatch between time of generation and time of need for wind and solar can be solved by expanding the geographic lattice and connecting to adjacent power grids. Hopefully, the wind will be blowing in area A, when it isn't in area B, and vice versa. But remember the reserve margin in each – and why it is required.

There are two major electric power grids in North America and three lesser but still substantial ones. Each grid has phase-synchronized 60 hertz (cycles per second) power within its own grid. But from grid-to-grid, although the frequency is still 60 hertz, they are not phase- synchronized and thus must undergo phase-synchronization at any interconnection with another grid. This involves converting (rectifying) the incoming power (side 1) to direct current and then inverting it back to alternating current that is phase-synchronized to side 2. Often the voltage is changed at that point. This all requires extra equipment (transformers, rectifiers, buses, inverters (or generators), more transformers, switchgear, and double some of this for a two-way interconnection capable of handling the large power load of the interconnection. Thus, such interconnections are not cheap compared to an ordinary interconnection between phase- synchronized systems within a synchronized grid. After all this expense, the interconnection is capable of transferring electricity from one grid to another, but only as much power as that interconnection can handle, which would be a small fraction of the total power required by the whole grid power demand.

Maybe the wind is blowing in grid zone A when it isn't in zone B, but does A have enough spare capacity to be able to help out B? Remember, we're talking day to day. So for A to be able to cooperate with B someone is going to have to build much more excess capacity in order to supply the needs of both A and B at the same time. The reserve margin is going to have to be materially increased. But by how much? Engineers are going to have to make assumptions. How long will the wind not blow in zone B? How much of zone B's wind power will be affected? So, after making assumptions, we begin adding considerable cost in the form of generating capacity in a place we hope won't run out of wind at the same time that we run out of wind. Determining the degree of interconnection that will be needed involves making huge assumptions about the amount of power to be produced by various generating sources (wind, solar, fossil, hydro, nuclear, etc.). Next, we'll need to assess the amount of excess capacity we'll need to "help our neighbors", what kind of energy source that is going to be, and who is going to pay for it. The answers to these questions keep changing as the degree of unreliable (wind and solar) increases and the reliable fossil fuel base load decreases. This almost makes the whole enterprise an impossible task to accomplish. One thing is known however, as the wind and solar penetrate further and further into the base load the cost will go up and up and up. And that is the scary part of the whole concept of increasing wind and solar power generation sources to a point beyond their being a fringe supplement.

Appendix F: Solar Proximity Magnetic and Radiational Effects

The earth's temperature is currently near the terminus of its more-or-less 10,000-year solar *proximity*-caused "genial platform" (its 10,000-year terminal plateau part of the orbital 100,000-year cycle shown to the far right in graphic AG.1). In addition, it is nearing the end of the current solar dual-magnetic wave cycle. This means the earth's surface central temperature trend-line will soon begin to bend downward as the sun's natural solar dual-magnetic wave cycle continues onward – forcing the earth to begin to cool.

The orbital (100,000-year) cycle is much longer and slower than the solar dual-magnetic wave cycle (which is multiple-hundreds to a thousand or so years) and will take perhaps a thousand to five thousand or more years to incur a one-degree Celsius drop in global temperature. Thus, this orbital cycle, while very important over long periods of time in explaining the significant (12°C) excursion we see over tens of thousands of years from ice-core proxies, does not have an appreciable effect on short term, human-measurement time frame global temperature.

The differential rotation of the sun's plasma-viscous mass sections apparently also causes a good deal of mixing of the interior constituents of the sun that, in turn, causes various solar surface occurrences that we witness as sun-spots, solar-flares, coronal-mass-ejections, coronal-holes, prominences, and field-lines – as well as the ionic solar wind. These solar surface occurrences contribute to the short-term changes in solar heat radiation. The longer (mid-term) time frame changing solar magnetic field due to the solar dual-magnetic wave cycle has a significant influence on the earth's sustained cloudiness over a multi-hundred-year time frame and thus on the reflection by clouds of solar heat radiation – and therefore on global surface temperature.

NASA, using a fleet of more than 20 different kinds of satellites (called the Heliophysics System Observatory) monitors the processes at work in the sun/earth relationship throughout the "near-earth space environment". The ionic solar wind is particularly studied because it can have a nuclear radiation effect on astronauts as they ascend through the atmosphere and out beyond the protection of the earth's magnetic shield.

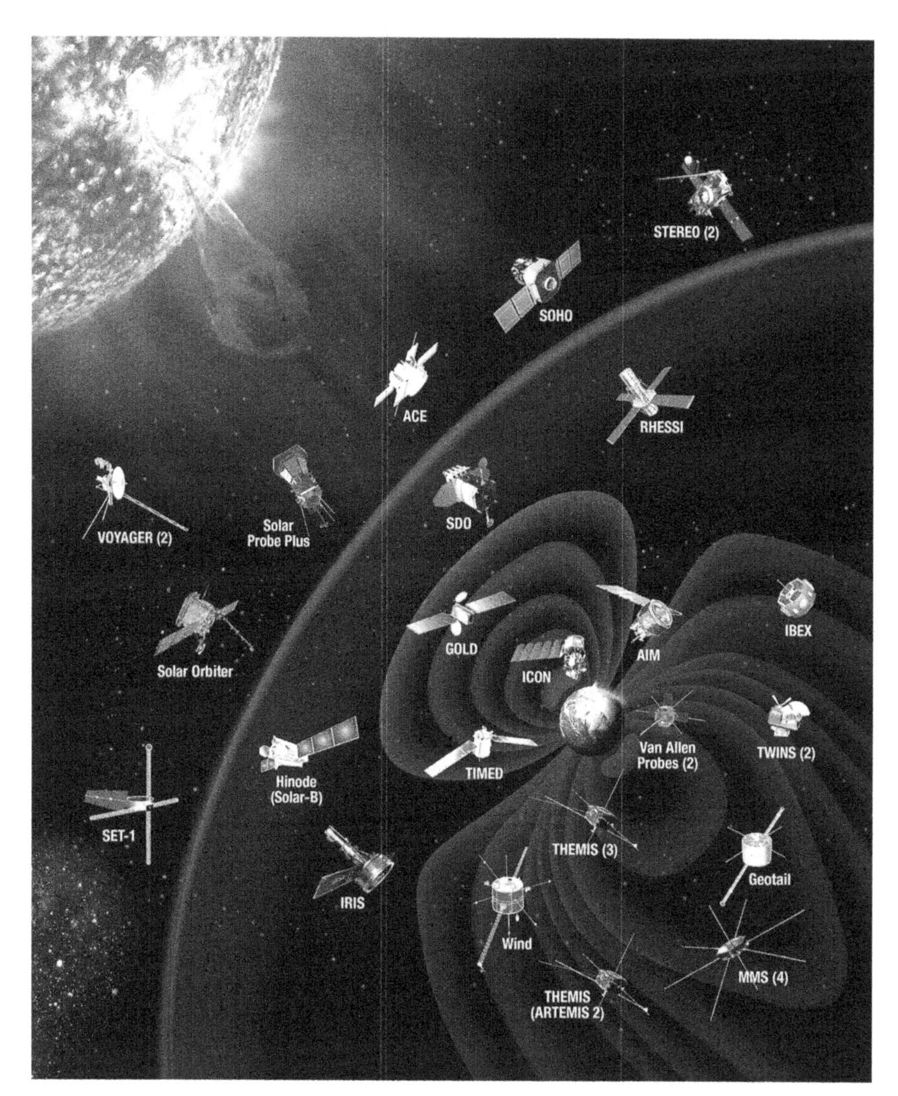

Graphic AF.1. **Heliophysics System Observatory.** NASA[236]. Showing: the sun – emitting its particulate "ionic solar wind"; the earth – with its protective "magnetic shield"; and various satellites monitoring the processes at work throughout the near-earth space environment.

Appendix G: The Earth's Global Warming/Climate Change Paradigm.

The following numbered paragraphs correspond to the numbered paragraphs in graphic 16.2 (the matrix showing factors influencing earth's temperature and climate).

1) The *proximity* of the earth to the sun is the most important factor in controlling global surface temperature. But the earth/sun distance change (and resulting temperature change) is extremely slow – taking up to thousands of years to show detectable temperature change and about 100,000 years to complete a full cycle. It is so slow as to be imperceptible to our human life-time observations. But it does show up in long-term proxies such as deep ice-core, marine sediment core, and cave speleothem data.

The distance between the sun and earth is basically set by the interaction of the many momentum, inertial, gravitational, Coriolis, centrifugal and centripetal forces involved that regulate the earth's complex relationship to the other bodies in the solar system. Temperature excursions in the range of 12°C or more over the full 100,000-year Milankovitch cycle were common in the past and can be expected in the future.

The third and fourth Milankovitch cycles caused by the "wobble" gravitational pull of Jupiter and Saturn on the earth (as well as the pull of the lesser planets) are of less (but still significant) gravitational influence on the primary gravitational cycle. They cause the shorter term (but not yet completely characterized) proximity changes between earth and sun that show as the major (taking place over thousands to tens of thousands of years) temperature spikes and prongs in earth's global surface temperature between the genial (maximum) and glacial (minimum) extremes (see graphic AG.1). These are termed the stadials and interstadials by scientists studying their characteristics.

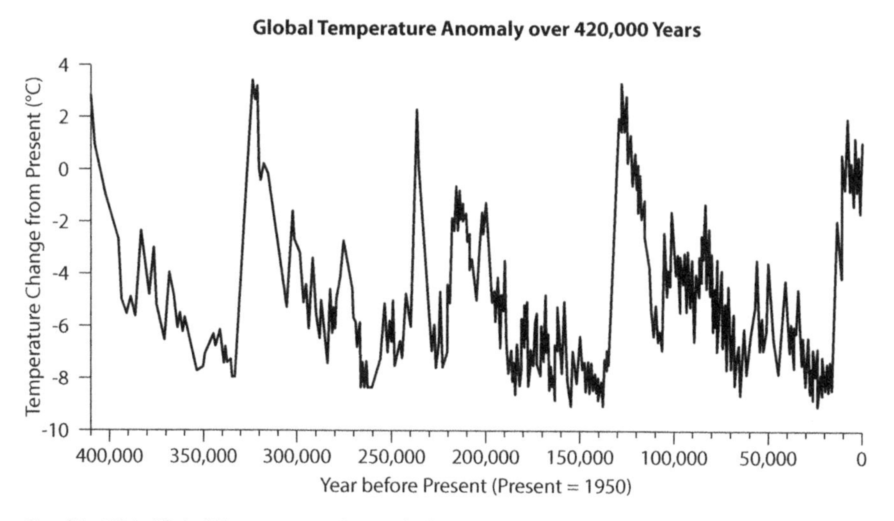

Graphic AG.1: **Global Temperature Anomaly Over 420,000 years.** J.R. Petit[237]. (Repeat from Chapter 1), This chart shows temperature for the past 420,000 years and also the improved resolution of the testing technique as time progresses. [Time →].

In graphic AG1, the lesser temperature spikes and prongs (over multiple-hundreds of years) are best explained by the solar dual-magnetic wave cycles and are superimposed on both the gravitationally caused stadials (over multiple-thousands to tens of thousands of years) and the basic, gravitationally caused 100,000-year cycle.

The temperature excursion spikes shown in graphic AG.2, also show the solar dual- magnetic wave cycle on another scale.

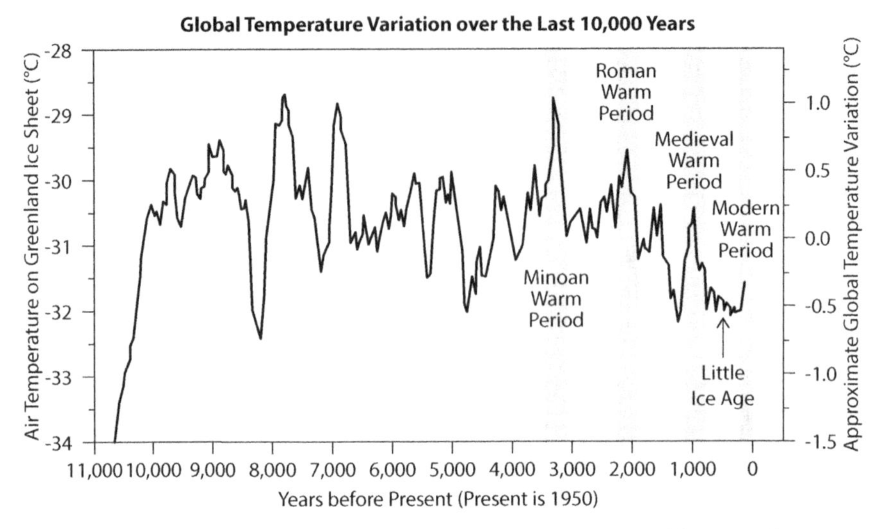

Graphic AG.2: **Global Temperature Variation last 10,000 years.** R.B. Alley[238]. (Repeat from Chapter 10). This chart shows global temperature variation caused by solar dual-magnetic wave cycles over a 10,000-year period. A more stretched-out version of the final "upswing" in this graphic is shown in the "Midlands" temperature record, graphic AG.3 [Time →].

Another interesting feature of Graphic AG.2 is that it apparently shows a long-term, orbital (proximity) cycle "going over the top" of the global temperature curve with a superimposed series of much shorter mid-term time frame magnetically induced cycles (which are themselves of varying extent). Thus, the ultimate temperature "peak" of the proximity cycle occurs somewhere along this 10,000-year "genial plateau" and it's difficult to tell just where – but it definitely seems to have peaked by the Minoan warm period some 3200 years ago (and perhaps even earlier somewhere around the 7800-year ago peak. After that it seems to decline toward its first lower "stadial plateau". However, the solar dual-magnetic wave cycles of the sun are superimposed and cause much shorter-term "ups" and "downs" on what is fundamentally a declining long-term temperature trend caused by the changing proximity between the earth and sun.

These revelations go a long way toward explaining the complex multiple factors that are causing earth's temperature to change in different time frames.

2) The earth's *changing cloud-cover* impacts the *reflectivity* of incoming solar heat and is influenced by the changing magnetic shield surrounding the earth which is highly affected by the solar dual-magnetic wave cycle. This, in turn, influences solar wind (ionic) as well as cosmic ray bombardment of earth's atmosphere creating particulate showers and subsequent cloud formation. The resulting cloud-caused change in earth's reflectivity to incoming solar heat causes most of the temperature spikes in the multi-hundreds-of-years, time frame shown in graphic AG.2. – including the one we are now in, shown both by the central trend line in graphic AG.3, and on a more compressed time scale, by the final prong upswing (to the right) in graphic AG.2. (which only shows to 1950 and must be extended to show todays temperature).

3) Changing *total solar irradiance* is most noticed as a short-term (up to a decade or so) period and is the most minor of the solar influences on earth's changing global surface temperature. Such temperature excursions are short and rarely exceed 1° to 3° C or so, and show up as some of the minor variations (background) in graphic AG.3.

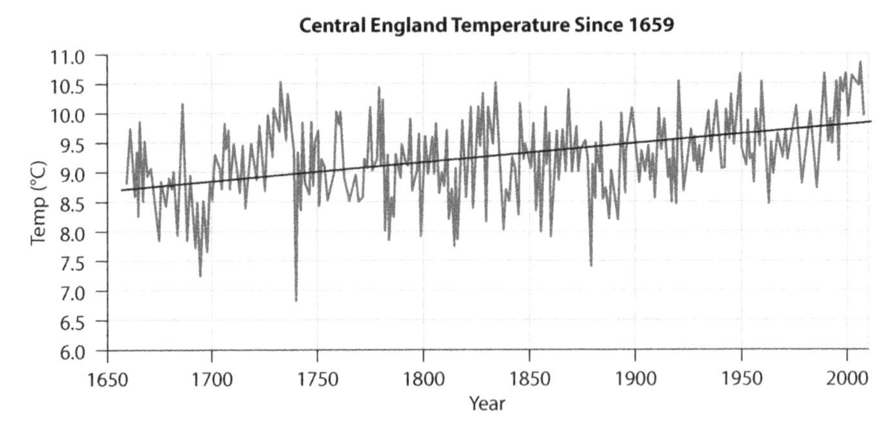

Graphic AG.3: **The Midlands of England.** G. Manley et al[239]. (Shown earlier in Chapter 1 – with added short-term uptick trend lines as well as this overall trend line). This is a more stretched-out version of the final upswing (to the right) shown in graphic AG.2. [Time →].

4) Oceanic and atmospheric colliding currents and overturning thermal layers can cause weather and short-term climate perturbations and weather extremes that are sometimes disturbing to humans but have no lasting overall global-heat-balance effect. They are indicative of weather and short-term climate variability and can last up to a decade or more. Some of the multi-decadal temperature variations such as the North Atlantic Oscillation and the various "see-saws" are probably in this category as well – although these multidecadal variations have not yet been well enough characterized to be fully understood.[240]

5) Human addition of carbon dioxide and other greenhouse gases to the atmosphere has been made (to say the least) into something significant in the minds of many people but essentially has only a trivial effect on

our global surface temperature and climate. It is, as one eloquent human observer, William Shakespeare, chronicled, "*Much ado about nothing*"[241]. The entire fossil fuel era will probably last less than a thousand years and will simply fade away as the fossil reserves are progressively depleted. Due to misinterpretations in the assumptions made when assessing the applicable physics, the influence of changing atmospheric carbon dioxide on global temperature is only trivial – not consequential as atmospheric climate scientists had led themselves to believe (You can't blame them for trying to make this conclusion that human-caused greenhouse gases were responsible for global warming into something since they had no other clearly-visible explanation available to them at that time).

Changing total solar irradiance [number (3) in the above list of influencing factors] is the only one of the solar effects that was considered by atmospheric inner-core climate scientists when they stated that "Changes in global temperature caused by the sun are not powerful enough to cause the changes in global surface temperature we are witnessing". Not enough was known about the sun at that time and a few key scientists misinterpreted the information that was available when they were forced to take a position concerning the cause of global warming. Under pressure for an answer, they drew a very hasty and uncertain (but fallacious) conclusion about the addition of CO_2 to the atmosphere being the culprit, which has grown into the unfortunate massive global warming and climate change illusion we have today.

The changing solar dual-magnetic wave phase progression cycle influence on earth's cloud cover is becoming better understood and is gaining credibility as an important component of mid-term time frame global surface temperature variation and climate change. It best describes many of the previously unexplained and misunderstood earth surface temperature changes that show on various proxies (as shown on graphic AG.2).

Acknowledgements

We wish to give particular recognition to Thomas Henry Huxley (1825 to 1895), who encouraged legitimate skepticism in the face of a strong but unverified conventional wisdom.

and

with special kudos to true scientists Antero Ollila, Hermann Harde, William Happer, Roy Spencer, Willie Soon, and Richard Lindzen who followed the science instead of the herd.

Over and above the credits cited, the following people had an impact on this book:

Kevin Bodge; Katherine Brennan; Ann Gurley; Syd Gurley; Don Henry; James R. Houston; Rody Johnson; the late Don Lindell; Paul Moyer; Bailey Pryor; Lee Pryor; Mardy Pryor; Ted Pryor; Vernon Rauch; Julie Smith, Julie Studt; L.S. Rothman; John Tierney.

Above all, thanks to:

- Our copy editor (and grammatical conscience) "Elsa" – with whom we had many excruciating, thought-provoking, crypt-esthesian debates (and who, much of the time, knew best.)

- Our very talented graphics expert Dan Swanson of Van-garde Imagery who took so many diverse graphs (and a few pictures) and made them look consistent and readable.
- Our specially accomplished indexing maven, Noalani Terry.
- And finally, our interior and exterior layout cognoscente Darlene Swanson of Van-garde Imagery who made the whole thing look like a book.

About the Author

Edward Rouse Pryor studies, lectures, and writes about the history of science and technology. As an Atlantic coastal resident, he became particularly interested in the basic causes of sea level rise (and thus in global warming) almost three decades ago.

He received his bachelor's degree in electrical engineering from the University of Virginia. While studying graduate geophysics at Columbia University he assisted Bruce Heezen and Marie Tharp in mapping the sea floor of parts of the Atlantic Ocean and the Red Sea.

He has worked in a variety of engineering, planning, and management positions at the Virginia Electric Power Company, the Westinghouse Electric Corporation, and the Electric Boat Division of General Dynamics, where he held positions of chief of manufacturing engineering, manager of industrial engineering, and manager of program planning and control. As president of the Stonington Product Development Company he participated in the development of an electric automobile. He holds one U.S. patent.

Throughout life, he has maintained a continuing interest in and study of the history of science and technology. He has a significant collection of embryonic scientific textbooks and handbooks dating back to the early 1800s.

His keen interest in sea level rise led him to accept the chairmanship of the Flood and Erosion Control Board of the coastal town of Groton, Connecticut. Later, for many years he was the chairman of the Beach and Dune Preservation Committee at Ocean Village, a condominium community on the Atlantic Coast in Florida.

He is a member of the American Geophysical Union, the American Association for the Advancement of Science, and the American Shore and Beach Preservation Association.

In 2009 the scientific honor society Sigma Xi published his *Man and Global Warming* as a part of its collection of white papers on The Year of Energy.

In October 2012, he presented a paper on anthropogenic sea level rise to the American Shore and Beach Preservation Association's technical conference in San Diego, California.

He was a tutor of mathematics while in college. He taught electronics while in the U.S. Naval Reserve. More recently, he has taught courses on *climate change, sea level rise*, the *Industrial Revolution*, and *our energy future,* at the Fielden Institute of Indian River State College in Florida. He has lectured general audiences in these same subjects.

As a citizen conservationist, he was a founding director of both the Avalonia Land Conservancy and the Groton Open Space Association, in Connecticut.

End Notes including: References, Citations, Credits, and Notes

THE *CITATION-OF-AUTHORITY* PROCESS IN the field of anthropogenic global warming and climate change is imperfect (as was pointed out during the Climategate inquiry). Apparently, commonly held beliefs passed peer review when a scrupulous examination would have revealed the flaws. But, with such a strong common preconception, that simply did not always happen.

But that's not to say that there is not a tremendous amount of honest, straightforward science that can be used in studying the subject. There is. This author perused hundreds of scientific papers and articles in an attempt to cover the field impartially. Some of these are cited in this End Notes section.

Since this book is for both general and scientific audiences, a balance of specific references is provided. There are not too many to be burdensome, but enough to assist in further inquiry by an interested and curious reader.

Below we present selected references, citations, credits, and notes as they are discussed in the book. In addition, there is a tremendous body of information in the literature that we do not cite because to do so would be too unwieldly. Thus, this material must be considered to be from the almost unlimited, vast general body of climate change literature. The amount of literature available on this topic is truly unbelievable. With $50 billion spent on "climate research" by the turn of the twenty-first century and perhaps much more than another $50 billion since, you can imagine how much scientific literature that is.

We hope that there is enough cited for you to find what you are looking for.

Endnotes

Introduction

1 Fourth National Climate Assessment Report, vol. 2 – a quadrennial review by the U.S. Global Change Research Program, a consortium of thirteen applicable U.S. government agencies. (November 23, 2018).

2 Wallace-Wells, D., *"The Uninhabitable Earth"*, New York Magazine, (July 9, 2017).

Chapter 1

3 Graphic 1.1. **The Midlands of England**. Source data: Manley, G., "*Central England temperatures: monthly means 1659 to 1973*", Quarterly Journal of the Royal Meteorological Society, vol. 100, (1974); additional compilations by the National Center for Atmospheric Research in Boulder Colorado and the Hadley Center in Exeter, England; plus added trend lines by the author. Final image by publisher (2020).

4 Graphic 1.2. **Global Temperature Anomaly Over 420,000** Years. Source: Petit, J.R., et al, *"Climate and Atmospheric History of the past 420,000 years from the Vostok Ice Core Antarctica"*, Nature, vol. 399, (3 June, 1999). Adapted from Philippe Rekacewicz, UNEP/GRID-Arendal. Final image by publisher (2020).

5 Ibid.

6 IPCC Assessment Reports 1-5. Intergovernmental Panel on Climate Change. (1990; 1995; 2001; 2007; 2013,14)

7 Hansen, J., "*Global Warming Has Begun.*" New York Times, front page, (January 23, 1988).

8 Kirby, A., "*Human effect on climate 'beyond doubt'.*" BBC News, front page, (22 January, 2001).

9 Carter, Jimmy, POTUS, National Climate Programs Act, signed: (1979).

10 IPCC Third Assessment Report, (January, 2001).

11 Graphic 1.3. **Radiative Forcing,** adapted from McInnes, Leland, GNU Free Documentation License, version 1.2; C.C. by S.A. 3.0. (c 2007).Indicator line added by author. Final image by publisher (2020).

12 Kirby, A., (2001).

13 IPCC Third Assessment Report, (January, 2001).

14 U.S. Government Accountability Office, (2015).

15 Citizens Climate Lobby, formed by Marshall Saunders in 2007; Cover: Climate/Public Citizen is a movement to make the public aware of the seriousness of human role in climate change; Congress Watch, another lobby group.

16 Schlesinger, M. and Andronova, N., "*Climate sensitivity: Uncertainty and Learning,*" Proceedings of the World Climate Conference, Moscow, (29 September to 3 October 2003

17 National Academies Press, Open Book 10139, Summary, National Science Foundation, (2001).

18 Soon, W.W., "*Variable solar irradiance as a plausible agent for multidecadal variations in the Arctic-wide surface air temperature record of the past 130 years*", Geophysical Research Letters 32: L 16712 (2005).

Chapter 2

19 Fourier, J., "*Memoirs of the Royal Academy of Sciences of the Institute of France*", pp. 569-604, (1827).

20 Tyndall, J., "*Contributions to Molecular Physics in the Domain of Radiant Heat*" [book], (1872).

21 IPCC Third Assessment Report, (January, 2001).

22 Arrhenius, S., "*On the Influence of Carbonic Acid in the Air upon the Temperature of the Ground*" Philosophical Magazine and Journal of Science of London, Edinburgh, and Dublin. vol. 41. no. 251. [Fifth series]. (April 1896).

23 Van Nostrand Scientific Encyclopedia, Fifth Edition, (1976).

24 Milankovitch, Milutin, "*Theorie Mathmatique des Phenomenes Thermiques par la Radiation Solaire*". Gauthiier- Villars, Paris. (1920).

25 Note: Recently a postulated Fifth Milankovitch-type cycle has been proposed: Surface geologists have observed an apparent overall 405,000 year (or 413,000 year) "megacycle". These cycles are not well defined theoretically but could exist as resonances or beat-frequencies from the interaction of some of the basic Milankovitch cycles. There is some observational evidence of this cycle from Chinese, Mediterranean, and Southwest U.S. cave speleothems (stalactites) as well as petrified wood from Arizona.

26 Hettner, G., "*Uber das Ultrarote Absorptionsspektrum des Wasserdampfes*", Annalen der Physik (Leipzig), series 4, vol. 55 (6), pp. 476-497 (1918).

27 Simpson, G.C., "*Further Studies in Terrestrial Radiation*", Memoirs of the Royal Meteorological Society 3 (21), pp. 1-26, (1928).

28 Plass, G.N., "*The Carbon Dioxide Theory of Climate Change*", Tellus VIII, 2, (1956).

29 Revelle, Roger, "*The Role of the Oceans*", Saturday Review, pg. 41 (7 May, 1966).

30 Gore, Al., An Inconvenient Truth, [book] Rodale Books, (2006)

31 Graphic 2.1. **Global Surface Temperature Since 1880.** Source: NASA/GISS. Credit: justfacts.com. Vertical indicator line added by author. Note: Not only NASA, but Met Office Hadley Center/CRU (U.K.), NOAA National Climate Center (Colorado), and the Japanese Meteorological Agency all show nearly identical results.

32 Graphic 2.2. Historical graph. **Late 1950s Through Early 1970s "Keeling Curve" of Atmospheric CO_2.** Source: Charles David Keeling, Scripps Institution of Oceanography. CC by S.A. (1972), Wikimedia.

33 Charney, J.G., et al, "Carbon Dioxide and Climate: A Scientific Assessment", Ad Hoc Study Group on Carbon Dioxide and Climate, (Charney Panel) National Academy of Sciences, [report] (July 23-27, 1979).

34 Carter, Jimmy, (1979).

35 IPCC Third Assessment Report, (2001).

36 Eddy, J., "*The Maunder Minimum*", Science, vol.192, number 4245, (18 June 1976).

37 Graphic 2.3. **Yearly Sunspot Count**, Source: NASA, Marshall Space Flight Center. Wikimedia (2020)

38 Hayes, J.D., Imbrie, J, and Shackleton, N.J., "*Variations in the Earth's Orbit: Pacemaker of the Ice Ages*", Science, vol. 194 number 4270, pp. 1121-1132, (10 Dec. 1976).

39 Kukla, G., "*Loess Stratigraphy of Central Europe…*" [book] Moutan Publishers, The Hague, pp 99-188 (1975).

40 Mesolella, R.K. et al, "*The Astronomical Theory of Climate Change: Barbados Data*" The Journal of Geology 77, no.3, (May, 1969).

41 Broecker, W. et al, "*Milankovitch hypothesis supported by precise dating of coral reefs and deep sea sediments*", Science, 159, pp 297-300, (1968)

42 Hoos, Ida R., Lecture, (1972) [attributed].

43 Larson, E., "*Isaac's Storm*" [book], Vintage Books, (2000).

44 Marx, R., "*Shipwrecks in Florida Waters*", [booklet] Mickler House, publishers, (1985).

45 Larson, E., (2000).

46 The Editors of Encyclopaedia Britannica, Encyclopaedia Britannica, (2018), Wikimedia.

47 Larson, E., (2000).

48 Larson, E., (2000).

Chapter 3

49 Kolbert, Elizabeth, "*Field Notes from a Catastrophe*", [book] Bloomsbury USA, (2006).

50 United Nations Framework Convention on Climate Change [UNFCCC] charter statement (1992).

51 IPCC First Assessment Report, (1990). [FAR].

52 Chikita, K. and Yamada, T., "*Sedimentary effects on the expansion of a Himalayan Lake*", Global and Planetary Change **28** (1-4) 23-24, (2001).

53 Ali, A., "*Vulnerability of Bangladesh Coastal Region to Climate Change*", Space Research and Remote Sensing Organization of Bangladesh [SPARRSO], (2000).

54 Kiehl, J.T. and Trenberth, K.E., "*Earth's annual mean energy budget*", Bulletin of the American Meteorological Society, vol. 78, no. 2, (February 1997). [K&T 97].

55 Kerr, R.A., "*Rising Global Temperature, Rising Uncertainty*", Science, Vol. 292, no. 5515, (13 April, 2001).

56 Stubenrauch, C.J., et al, "*Assessment of global cloud datasets from satellites…*", Bulletin of the American Meteorological Society 94 (7), (2013).

57 [K&T 97].

58 Pryor, E.R., "*Anthropogenic Sea Level Rise*" [lecture] – American Shore and Beach Preservation Association, San Diego, California, (October, 2012).

59 Miskolczi, F., "*The stable stationary value of the earth's global average atmospheric Plank-weighted greenhouse gas optical thickness*", Energy and Environment, Special Issue: Paradigms in Climate Research, v. 21, number 4, (August, 2010).

60 Markoff, J., "*A climate modeling strategy that won't hurt the climate*", New York Times, (May 11, 2015).

61 [K&T 97].

62 Graphic 3.1. **The 2001 - IPCC Temperature "Hockey Stick"**. Data source, IPCC Third Assessment Report, (January 2001). Final image by publisher (2020).

63 Petit, J.R., et al, "*Climate and atmospheric history of the past 420,000 years from the Vostok ice core in Antarctica*", Nature, v. 399, pgs. 429-436 [after], (3 June, 1999).

64 Graphic 3.2. **Vostok Ice-Core Data.** Source: Petit, J.R., et al, "*Climate and atmospheric history of the past 420,000 years from the Vostok ice core in Antarctica*", Nature, v. 399, pgs. 429-436 (3 June, 1999). Adapted from Philippe Rekacewicz, UNEP/GRID-Arendal. Final image by publisher, (2020).

65 [K&T 97].

66 Charney, J., (1979).

67 IPCC Third Assessment Report, (2001).

68 [K&T 97].

69 IPCC Third Assessment Report, [Michael Mann, Penn State University], (2001).

70 Petit, J.R., (1999).

71 IPCC Third Assessment Repot, preliminary draft, (2000).

72 Kirby, A., (2001).

Chapter 4

73 Kerr, R.A., (2001)

74 Horner, C., "*Red Hot Lies*", [book] (quoting Time Magazine Managing Editor Richard Stengel comments made on NSNBC), Regnery Publishing, (2008).

75 Gore, Al, "*An Inconvenient Truth*", [book] Rodale Books, (2006).

76 Documentary film: "*An inconvenient Truth*", [motion picture] starring Al Gore. Directed by Davis Guggenheim. (2006).

77 Nobel Peace Prize, 2007.

78 US Environmental Protection Agency, "*Endangerment Finding*", 2009. This became the legal foundation for climate action by the federal government. It required the EPA to take action to curb emissions of greenhouse gases that endanger public health by contributing to climate change.

79 [K&T '97].

80 IPCC Third Assessment Report, January, 2001.

81 IPCC Fourth Assessment Report, 2007.

Chapter 5

82 [K&T '97].

83 Polo, Marco, "*The Description [Marvels] of the World/The Travels of Marco Polo*", Venice, (c 1299).

84 Pryor, E.R. Personal observation during the exploration of the state of Rhode Island's early industrial sites (1975).

85 Graphic 5.1. **Dams in the State of Rhode Island.** Historical map. Data – U.S. Geological Survey. Image – State of Rhode Island Department of Environmental Management.

86 Rivard, P.E., "*Samuel Slater Father of American Manufactures*", [essay], Slater Mill Historic Site, (1974).

87 Segar, G. and Salomon, B. "*Water Power Revisited*" [booklet], Rhode Island Committee for the Humanities, (1980).

88 Graphic 5.2. **The Corliss Four-Valve Steam Engine at the 1876 Philadelphia Exhibition.** Historical etching. "*The Engineer's Handy Book*", (1884) Public domain. Wikimedia Commons.

89 A compilation from Wikimedia Commons.

90 Graphic 5.3. **De Witt Clinton Steam Train**. Historical photograph. Creative Commons Attribution-Share Alike License-3. Henry Ford Museum, Detroit, Michigan, Wikimedia Commons.

91 Graphic 5.4. **Winton Motor Carriage.** Winton Motor Carriage Company. Historical first auto advertisement, (1898). Public Domain, Wikimedia Commons.

92 Public discussion with Westinghouse Electric Corporation executives.

93 Public discussion with mine executives, Arch Coal, Black Thunder Mine, Wright, Wyoming, 2003.

94 National Gypsum Company: Apollo Beach, FL; Shippingport, PA.; Westwego, LA; and Mount Holly, NC., all have tandem artificial gypsum wallboard manufacturing facilities associated with electric generating plants.

Chapter 6

95 U.S. Energy Information Administration.

96 Happer, W., "*The Truth about Greenhouse Gases*", First Things, June/July 2011.

97 IPCC Third Assessment Report, January 2001.

98 McEntee, C. (CEO of the American Geophysical Union), public address (29 January, 2014).

99 Ibid.

100 Schwarzenegger, Arnold, (Governor of California), World Environmental Day speech, (June 1, 2005).

101 Kerr, R. A. (2001).

102 Kirby, A. (2001).

103 Tierney, J. "*Informational Cascade*", New York Times, (October 9, 2007).

104 Schopenhauer, A., "*The Art of Controversy*" chapter 3 (posthumous paper, 1896) as reported by John Tierney in the New York Times, (Oct 9, 2007).

105 Janus, Irving L., Victims of Groupthink [book] Houghton Mifflin Company (1972).

106 Sondergard, S.E., Climate Balance, [book] Third Edition, Yorkshire Publishing (2016).

107 Tierney, J. (2007)

Chapter 7

108 Rio Climate Summit, conference convened by the United Nations Framework Convention on Climate Change [UNFCCC] in Brazil, (1992)

109 Kyoto Protocol, conference convened by the UNFCCC in Japan, (1997).

110 Stout, D., "*Supreme Court Backs States on Clean Air Act and Greenhouse Gases*", N. Y. Times, (April 2, 2007).

111 EPA, "*Endangerment and Cause…*" Federal Register, part V, chapter 1, (Dec. 15, 2009).

112 "*Clean Power Plan*", U.S. Environmental Protection Agency, (June, 2014).

113 Moore, C., et al, "*Advanced Exploration and Technology Concepts…*", Seasonal and Discovery article # 42047, AAPG data pages, On Line Journal for E & P Geoscientists (2017).

114 Copenhagen Summit, UNFCCC, in Denmark (2009).

115 Paris Climate Agreement, UNFCCC in France (2015).

116 "*Power Plant Rejected Over Carbon Dioxide for First Time*", Washington Post, (Oct. 20, 2007).

117 Korte, G. and Jackson, D. "*Obama administration rejects Keystone pipeline*" USA TODAY, (Nov. 6, 2015). [This was the controversial proposed Keystone XL pipeline fourth extension through Nebraska].

118 "Clean Power Plan", U.S. Environmental Protection Agency, (June, 2014).

119 Executive Order on Energy Independence [E.O. 13783] signed by POTUS, D. Trump, (March 28, 2017).

120 Shear, M.D., "*Trump Will Withdraw U.S. from Paris Climate Agreement*", New York Times, (June 1, 2017).

Chapter 8

121 Sharer, K., former CEO of biotech giant Amgen. Quoted by Bret Stephens, The New York Times, April 18, 2020.

122 Happer, W., (2011).

123 Allen, L.H. Jr, "*Effects of Increasing Carbon Dioxide Levels and Climate Change on Plant Growth…*" (A comprehensive study compiled from some 92 scientific papers), Managing Water Resources in the West Under Conditions of Climate Uncertainty [book] Chapter 7, National Academies Press A, Proceedings. (1991). A more recent satellite study by a team of 32 authors from 24 institutions shows significant greening worldwide as a result of carbon dioxide greening, Journal: Nature Climate Change, (April 25, 2016).

124 Ibid.

125 Happer, W., (2011).

126 Graphic 8.1 **Carbon Dioxide Is Not an Atmospheric Pollutant.** Most source data: Happer, W. (2011). Matrix and image: Pryor, E. (2019).

127 Luthi, D., et al, EPICA Dome C Ice Core 800,000-year Carbon Dioxide Data (2008). IGBP pages/World Data Center for Paleoclimatology Data Contribution Series # 2008-055. NOAA/NCDC Paleoclimatology Program, Boulder CO, USA – accessed from the Carbon Dioxide Information Analysis Center, Oak Ridge National Laboratory, U. S. Department of Energy.

128 Kuhn, T., The Copernican Revolution [book], Cambridge: Harvard University Press. (1957)

129 Russel, B., History of Western Philosophy [book], Routledge. (1995).

130 Note: See Wikipedia entry under Classical Elements.

131 Wade, Lizzie, *"Feeding the Gods"*, Science Magazine, (22 June, 2018).

132 Significant Earthquake Database, U.S. Geological Survey.

133 "Chile asking for calm", Time Magazine, (4 July, 1960).

134 Delpech, M., "*Case of a wound to the right carotid artery*", (1825) Lancet 6 (73), Quoted in Carter, K, et al, Childhood Fever [book] (Feb 1, 2005).

135 Note: See Wikipedia entry under Phlogiston Theory.

136 McEntee, C., (2014)

Chapter 9

137 [K&T '97].

138 Harde, H., "*How much CO_2 really contributes to Global Warming? …*", Geophysical Research Abstracts, vol. 13, EGU 2011 – 4505 – 1, (2011).

139 [K&T 97].

140 Harde, H., "*Advanced Two-Layer Climate Model for the Assessment of Global Warming by CO_2*", Open Journal of Atmospheric and Climate Change, vol. 1, no. 3, pp 1-50 (2014).

141 Rothman, L.S. et al, "*The HITRAN 2008 molecular spectroscopic database*", Journal of Quantitative Spectroscopic Radiative Transfer 110, 533-572 (2009).

142 Harde, H., (2014)

143 Chen, L, "Cloud and Water Vapor Feedbacks..." Chinese Academy of Sciences, (12 July, 2013).

144 Idso, S.B., "*CO_2-induced global warming: A skeptic's view of potential climate change*", Climate Research, vol. 10, No. 1, (April 9, 1998).

145 Harde, H. (2011)

146 Lindzen, R. S., and Choi, Y.S., "*On the determination of climate feedback from ERBE data*". Geophysical Research Letters 36: 10.1029 (2009).

147 Lindzen, R.S. and Choi, Y.S. "*On the observational determination of climate sensitivity and its implications*". Asia-Pacific Journal of Atmospheric Science 47: 377-390. (2011).

148 IPCC Fourth Assessment Report, (2007).

149 Harde, H., (2014).

150 [K&T 97].

151 IPCC, Climate Change 2007 [Fourth Assessment Report]: The Physical Science Basis, Summary for Policymakers, page 2.

152 Wasdell, D., "*Arctic Dynamics*", Apollo-GAIA, (June 2013).

153 Note: This can be deduced by looking at Graphic 9.1 with the indicator line added by the author. (2019).

154 Graphic 9.1: **Atmospheric Infrared Absorption Bands.** © Rhode R.A. (2007), GNU Free Documentation License, version 1.2 and CCA-SA 3.0 unported, Wikimedia Commons (with superimposed alignment line by author, final image by publisher 2020).

155 Note: The small percentage of lower atmosphere over deserts has a lower water vapor content than the atmosphere over ocean or "green" terrestrial areas. But even so, it only takes a tiny amount of water vapor to cancel out most of the effect of the trace gases nitrous oxide and methane because their own concentration is so ultra-tiny in the atmosphere.

156 Graphic 9.2: **Hockey Stick Comparison.** Upper graph: IPCC First Assessment Report, (1990). Lower graph: Data source: Mann, M., IPCC Third Assessment Report, (2001).

157 Booker, C. "*Climate change: This is the worst scientific scandal of our generation*". The Telegraph (18 November, 2009); and Monbiot, G., "*Pretending*

the climate email leak isn't a crisis won't make it go away", The Guardian (29 November, 2009) are but two of a huge compilation.

158 Adam, David, *"Scientists cleared of malpractice in UEA's hacked emails inquiry"*, The Guardian, (14 April 2010).

159 Graphic 10.1 **Remains of Viking Church, Hvalsey, Greenland.** Historical photograph by: Ruttel, Frederik Cark Peter (1859-1915), Hvalsey, Greenland, Viking Church from the medieval warm period, Wikimedia Commons, Public Domain.

160 Graphic 10.2: **Historical Oil Painting by Abraham Hondius**: *"A Frost Fair on the Thames at Temple Stairs"* (1684), © Museum of London, Clamor-World, Wikimedia (CC by SA 4.0).

161 Suissman, B. "Climategate: A Veteran Meteorologist Exposes Global Warming Scam" [book], (April 2, 2010).

162 Graphic 10.3 **Global Temperature Variation, Last 10,000 Years.** Alley, R.B., Journal of Quaternary Science Reviews, 19:223-226. (Jan. 2000).

Chapter 11

163 Graphic 11.1. **Vostok Ice Core Data.** Historical Image by Monin, E., et al, *"Atmospheric CO_2 Concentrations over the last glacial termination."*, Science 291 (5501) pp. 112-114 (5 January, 2001).

164 IPCC Third Assessment Report, (January 2001).

165 IPCC Fourth Assessment Report (AR4), Climate Change 2007, Working Group 1, The Physical Science Basis. (2007).

166 Popkin, G., *"DOE unveils climate model in advance of global test"*, Science Magazine, vol. 360, issue 6388 pg.474, (May 4, 2018).

167 Graphic 11.2. **Temperature – the Midlands of England.** (Repeat from chapter 1).Data: Manley, G., *"Central England temperatures: monthly means 1659 to 1973"*, Quarterly Journal of the Royal Meteorological Society, vol. 100, (1974); additional compilations by the National Center for Atmospheric Research in Boulder Colorado and the Hadley Center in Exeter, England; plus added trend lines by the author. Final image by publisher (2020).

Chapter 12

168 Schellenberger, M. *"Unreliable Nature of Solar and Wind Makes Electricity More Expensive, New Study Finds"*, Forbes.com, (April 22, 2019).

169 Bryce, R. *"The Real Problem with Renewables"*, Forbes Magazine, (May 11, 2010)

170 Johnson, Rody, Chasing the Wind [book], University of Tennessee Press, (2014)

171 Morison, R., "*Britain has gone nine days without wind power*", Bloomberg News, (June 7, 2018).

172 Mountain, B. and Chang, C. "*South Australia has the highest power prices in the world,*" News.com.au, (9 August 2017).

173 Durwall, R., "*The Climate Change Act at Ten: History's most expensive virtue signal*", Global Warming Policy Forum, (November 2018).

174 Jacobson, M.Z. and Delucchi, M.A., "*A path to sustainable energy by 2030*" Scientific American, (Nov., 2009).

175 Ibid.

176 Shellenberger, M., "*If Solar and Wind are so Cheap, Why are they Making Electricity so Expensive?*" Forbes.com, (April 23, 2018).

177 Penn, Ivan, "*It Harnesses a River, Why Stop There?*", New York Times, (Aug. 4, 2018).

Chapter 13

178 Fourth National Climate Assessment, Vol. ll, Executive Summary, U.S. Global Change Research Program, Washington, DC, (November 2018) and as reported on and amplified by C. Davenport and K. Pierre-Louis in the New York Times, page A1 and A17, (November 24, 2018).

179 IPCC Fourth Assessment Report (2007), Detection and Attribution section.

180 Note: The IPCC often uses consumer friendly language: 66% = "likely"; 90% = "very likely"; 95% = "extremely likely".

181 IPCC Fifth Assessment Report. (2013-14).

Chapter 14

182 Note: Climate forcers are those qualities that cause the earth's surface temperature to change.

183 Graphic 14.1. **Radiative Forcing.** Mclnnes, Leland, GNU Free Documentation License, Version 1.2; C.C. by S.A. 3.0. (c. 2007). Final image by publisher (2020).

184 Graphic 14.2. **Greenhouse Gases versus Solar Forcing (As Claimed by IPCC)**. Derivation in Appendix B: Information from Leland Mclnnes, IPCC AR4, (2007), chart rendition by the author, (2020).

185 Graphic 14.3. **Graphic 14.2 Updated for Reduction of Inappropriate Greenhouse Gas Forcing and Addition of Appropriate Solar Forcing.** Information materially altered by E.R. Pryor (in accordance with logic shown in appendix B), from that shown by IPCC AR4. (2020).

186 See appendix B. for complete explanation.

187 Kerr, R.A. (2001)

Chapter 15

188 Zharkova, V.V, Shepherd, S.J., Zharkov, S.I., et al "*Oscillations of the baseline of solar magnetic field and solar irradiance on a millennial timescale*". Sci Rep 9, 9197 (2019) doi:10.1038/s41498-019-45585-3.

189 Zharkova, V., et al, "*Reinforcing a Double-Dynamo Model with Solar-Terrestrial Activity in the Past Three Millenia*", Proceedings of the International Astronomical Union, (May, 2017).

190 Popova, E., et al, "*Reinforcement of double dynamo waves as a source of solar activity and its prediction on millennium timescale*", American Geophysical Union, Fall General Assembly 2016, abstract id.SH43D-2588 (12/2016).

191 Tobias, S, M, "*The Solar Dynamo*" Philosophical Transactions of the Royal Society A. 360 (1801): 2741-2756 (2002).

192 Graphic 15.1. **The Three Section Sun**. Showing its three viscous-plasma sections rotating at slightly different velocities. Data – Tobias, S.M. (2002); Zharkova, V., et al (2017); Benomar, O., et al (2018); rendition – Pryor, E.R. © (2019).

193 Benomar, O., et al, "*Asteroseismic detection of latitudinal differential rotation in 13 Sun-like stars*), Science, vol. 361 issue 6408 (21 September 2018).

194 Graphic 15.2. **Global Temperature Variation Last 10,000 Years**. Repeat of graphic 10.3. Data: Alley, R.B., Journal of Quaternary Science Reviews, 19:223-226. (Jan. 2000).

195 Graphic 15.3 **The Midlands of England**. (Repeat from chapter 1). Data and rendition – Manley, G., "*Central England temperatures: monthly means 1659 to 1973*", Quarterly Journal of the Royal Meteorological Society, vol. 100, (1974); additional compilations by the National Center for Atmospheric Research in Boulder Colorado and the Hadley Center in Exeter, England; plus added trend lines by the author. Final image by publisher (2020).

196 Richard, K. "Papers Forecast Global Cooling". Notrickszone [scientific journal summary] (December 28, 2017).

197 Harde (2014).

198 Graphic 15.4 **Global Temperature Anomaly Over 420,000 years**. (Repeat from chapter 1). Petit, J.R., et al, "*Climate and Atmospheric History of the past 420,000 years from the Vostok Ice Core Antarctica*", Nature, vol. 399, (3 June, 1999). Adapted from Philippe Rekacewicz, UNEP/GRID-Arendal. Final image by publisher (2020).

Chapter 16

199 Keats, John, "*When I Have Fears That I May Cease to Be*" [poem] (1820), public domain.

200 Graphic 16.1 **Earth/Sun relationship Matrix.** Pryor, E., © (2019).

201 Note on clouds: Incoming solar irradiance is either reflected by, absorbed by, or transmitted through a cloud. The same is true of outgoing infrared radiation from the earth (although it is of a different frequency than the incoming radiation). Furthermore, clouds are at different elevations, are of different thicknesses, densities and shades. Thus, sustained changes to average cloudiness, has a complex, consequential effect on both reflectivity of incoming solar irradiance and retention of outgoing heat from the earth.

202 Kirby (2001).

203 Ollila, Antero, "*Cosmic Theories and Greenhouse Gases as Explanations of Global Warming*", Journal of Earth Sciences and Geotechnical Engineering, vol. 5, no. 4, pp 27-43 (2015).

204 Ibid.

205 Oreskes, N, "*The Scientific Consensus on Climate Change*", Science, vol. 306 Issue 5702, (03 Dec. 2004).

206 Nobel prize in Chemistry, (1948)

207 Graphic 16.2. **Factors Influencing Earth's Global Surface Temperature and Climate**. Matrix. Pryor, E. R. © (2019).

208 Note: The major long-term deep temperature excursions are best explained by the major proximity changes between earth and sun caused by the Milankovitch cycles, although a few solarists are exploring alternate long-term forces within the sun that may have influence.

Chapter 17

209 U.S. Office of War Information. "*Valley of the Tennessee*". Documentary Film, (1944).

210 Graphic 17.1. **World Population Growth Over the Last 12,000 Years.** Max Roser, Hannah Ritchie and Esteban Ortiz-Ospina (2019) - "*World Population Growth*". *Published online at OurWorldInData.org.* Retrieved from: 'https://ourworldindata.org/world-population-growth' [Online Resource]. CC by SA. (2020).

211 "*Juliana vs. the United States*", as reported by Steve Croft on "60 Minutes, CBS News, (March 3, 2019).

212 Ramzy, A. "*Students Across the World Are on Strike. Here's Why.*" New York Times, (March 15, 2019).

Chapter 18

213 Some say George Santayana actually said this.

214 Hubbert, M.K., "*Energy from Fossil Fuels*", Science, New Age; (1949)

215 Graphic E.1: Historical image of **Hubbert's 1949 Epoch of Fossil-Fuel Exploitation in Human History During the Period from 5000 Years Ago to 5000 Years in the Future**. Hubbert, M.K., *"Energy from Fossil Fuels"*, Science, New Age; (1949)

216 Graphic E.2: **Global Nuclear History.** Chart, from "*History of Nuclear Power in the world*". Data is from IAEA and EIA (2014) Creative Commons, CCO 1.0 Universal Public Domain Dedication, Wikimedia Commons, (25 August 2014).

217 Note: ITER stands for "International Thermonuclear Experimental Reactor" It is an eleven-nation scientific cooperative effort to develop a practical fusion reactor.

218 Note: DEMO is an acronym for Demonstration Power Station – the next generation of fusion power development following the ITER project.

219 Graphic AW.1, **Atmospheric Carbon Dioxide Content 1957- 2020**. Adapted from NOAA, Scripps Institution of Oceanography (2020).

220 New York Times, page B 1, April 14, 2020.

221 Voltaire, "*Candide*", (1759)

Appendices

222 Schmidt, G.A., "*Attribution of the present-day greenhouse effect*", Journal of Geophysical Research, 115, D20106, pg. 3, (16 October, 2010).

223 Graphic AA.1. **Sun and Earth Emission Curves,** adapted from J.C. Baez. Wikimedia commons.

224 Graphic AA.2: **Atmospheric Infrared Absorption Bands.** © Rhode R.A. (2007), GNU Free Documentation License, version 1.2 and CCA-SA 3.0 unported, Wikimedia Commons (with superimposed alignment line by author 2020).

225 Peixoto and Oort. *Physics of climate*, [scientific encyclopedia] pg. 93 (1992).

226 Graphic AA.3: **Water Vapor and CO_2 Infrared Absorption Spectra.** Data source: Simmon, R. NASA, Wikimedia Commons.

227 Harde, H. (2014)

228 Graphic AB.1: **Radiative Forcing.** McInnes, Leland, GNU Free Documentation License, Version 1.2; C.C. by S.A. 3.0. (c. 2007).

229 Graphic AB.2: **Radiative Forcing.** Forcing data from IPCC Third Assessment Report. Chart by Pryor, E.R., from data by IPCC, (2019).

230 Graphic AB.3: **Radiative Forcing. Corrected for Reduction of Inappropriate Greenhouse Gas Forcing and Addition of Appropriate Solar Forcing.** Information materially altered by E.R. Pryor (in accordance with logic shown in appendix B), from that shown by IPCC AR4. (2020).

231 Harde, H. (2014)

232 Parrenin, F., et al, *"Synchronous change of atmospheric CO2 and Antarctic temperature during the last deglacial warming,"* Science, (March 1, 2013).

233 Shakun, J.D., et al., *"Global warming preceded by increasing carbon dioxide concentrations during the last deglaciation"*, Nature Magazine, vol. 484, (5 April, 2012).

234 Eyre, B., et al *"Coral reefs will transition to net dissolving before end of century"*, Science 359/6378, (23 February, 2018)

235 Barkley, H., *"Scientists pinpoint how ocean acidification weakens coral skeletons"* Proceedings of the National Academy of Sciences, (January 29, 2018).

236 Graphic AF.1. **Heliophysics System Observatory**. Chart, NASA.

237 Graphic AG.1. **Global Temperature Anomaly over 420,000 years.** (Repeat from chapter 1). Petit, J.R., et al, *"Climate and Atmospheric History of the past 420,000 years from the Vostok Ice Core Antarctica"*, Nature, vol. 399, (3 June, 1999). Adapted from Philippe Rekacewicz, UNEP/GRID-Arendal. (2020).

238 Graphic AG.2. **Global Temperature Variation last 10,000 years.** Repeat from chapter 10. Alley, R.B., Journal of Quaternary Science Reviews, 19:223-226. (Jan. 2000).

239 Graphic AG.3. **The Midlands of England.** Repeat. Data: Manley, G, "*Central England temperatures: monthly means 1659 to 1973*", Quarterly Journal of the Royal Meteorological Society, vol. 100, (1974); additional compilations by the National Center for Atmospheric Research in Boulder Colorado and the Hadley Center in Exeter, England; plus an added trend line by the author. Final image by publisher (2020).

240 (a). Zarkova, V.V. et al "Reinforcing the double-dynamo model with solar-terrestrial activity in the last three millennia", Proceedings of the International Astronomical Union, (May 2017).

(b). Svensmark, J., et al, "The response of clouds and aerosols to cosmic ray decreases", Journal of Geophysical Research: Space Physics, (2016) [Two among many others].

241 Shakespeare, W., Play, c 1599.

Index

C

Q

R

S

Y

Made in the USA
Las Vegas, NV
10 August 2021